任务型语码转换式双语教学系列教材

总主编 刘玉彬 副总主编 杜元虎 总主审 段晓东

自动化与检测技术

AUTOMATION AND MEASUREMENT TECHNOLOGY

主 编 刘岩川 张 艳
副主编 陈晓云 赵凤强
主 审 田 森

大连理工大学出版社

图书在版编目(CIP)数据

自动化与检测技术 / 刘岩川，张艳主编. — 大连：
大连理工大学出版社，2014.6(2017.7重印)
任务型语码转换式双语教学系列教材
ISBN 978-7-5611-9121-7

Ⅰ.①自… Ⅱ.①刘… ②张… Ⅲ.①自动检测—双语教学—高等学校—教材—英、汉 Ⅳ.①TP274

中国版本图书馆 CIP 数据核字(2014)第 094332 号

大连理工大学出版社出版
地址：大连市软件园路80号 邮政编码：116023
发行：0411-84708842 邮购：0411-84708943 传真：0411-84701466
E-mail:dutp@dutp.cn URL:http://www.dutp.cn
大连理工印刷有限公司印刷 大连理工大学出版社发行

幅面尺寸:183mm×233mm 印张:13.75 字数:456千字
2014年6月第1版 2017年7月第2次印刷

责任编辑：邵 婉 责任校对：诗 宇
封面设计：波 朗

ISBN 978-7-5611-9121-7 定 价：28.00元

本书如有印装质量问题，请与我社发行部联系更换。

总序

2014年的初夏,我们为广大师生奉上这套"任务型语码转换式双语教学系列教材"。

"任务型语码转换式双语教学"是双语教学内涵建设的成果,主要由两大模块构成:课上,以不影响学科授课进度为前提,根据学生实际、专业特点、学年变化及社会需求等,适时适量地渗透英语专业语汇、语句、语段或语篇,"润物细无声"般地扩大学生专业语汇量,提高学生专业英语能力;课外,可向学生提供多种选择的"用中学"平台,如英语科技文献翻译、英语实验报告、英语学术论文、英语小论文、英语课程设计报告、模拟国际研讨会、英语辩论、工作室英语讨论会等,使学生的专业英语实践及应用达到一定频度和数量,激活英语与学科知识的相互渗透,培养学生用英语学习、科研、工作的能力及适应教育国际化和经济一体化的能力。

为保证"任务型语码转换式双语教学"有计划、系统、高效、科学地持续运行,减少教学的随意性和盲目性,方便师生的教与学,我们编写了这套"任务型语码转换式双语教学系列教材"。

本套教材的全部内容均采用汉英双语编写。

教材按专业组册,涵盖所有主干专业课和专业基础课,力求较为全面地反映各学科领域的知识体系。

分册教材编写以中文版课程教材为单位,即一门课为分册教材的一章,每章内容以中文版教材章节为序,每门课以一本中文教材为蓝本,兼顾其他同类教材内容,蓝本教材绝大部分是面向21世纪的国家规划教材。

教材的词汇短语部分,注意体现学科发展的新词、新语,同时考虑课程需求及专业特点,在不同程度上灵活渗透了各章节的重要概念、定义,概述了体现章节内容主旨的语句及语段。分册教材还编写了体现各自专业特点的渗透内容,如例题及解题方法,课程的发生、发展及前沿简介,图示,实验原理,合同文本,案例分析,法条,计算机操作错误提示等。

部分教材补充了中文教材未能体现的先进理论、先进工艺、先进材料或先进方法的核心内容,弥补了某些中文教材内容相对滞后的不足;部分教材概述了各自专业常用研究方法、最新研究成果及学术发展的趋势动态;部分教材还选择性地把编者的部分科研成果转化为教材内容,以期启发学生的创新思维,开阔学生的视野,丰富学生的知识结构,从教材角度支持学生参与科研活动。

本套教材大多数分册都编写了对"用中学"任务实施具有指导性的内容,应用性内容的设计及编写比例因专业而异。与专业紧密结合的应用性内容包括英语写作介绍,如英语实验报告写作、英语论文写作、英语论文摘要写作、英语产品、作品或项目的概要介绍写作等。应用性内容的编写旨在降低学生参与各种实践应用活动的难度,提高学生参与"用中学"活动的可实现性,帮助学生提高完成"用中学"任务的质量水平。

考虑学生英语写作和汉译英的方便,多数分册教材都编写了词汇与短语索引。

"任务型语码转换式双语教学系列教材"尚属尝试性首创,是多人辛勤耐心劳作的结果。尽管在编写过程中,我们一边使用一边修改,力求教材的实用性、知识性、先进性融为一体,希望教材能对学生专业语汇积累及专业资料阅读、英语写作、英汉互译能力的提高发挥作用;尽管编者在教材编写的同时也都在实践"任务型语码转换式双语教学",但由于我们缺乏经验,学识水平和占有资料有限,加上为使学生尽早使用教材,编写时间仓促,在教材内容编写、译文处理、分类体系等方面存在缺点、疏忽和失误,恳请各方专家和广大师生对本套教材提出批评和建议,以期再版时更加完善。

在教材的编写过程中,大量中外出版物中的内容给了我们重要启示和权威性的参考帮助,在此,我们谨向有关资料的编著者致以诚挚的谢意!

编 者

2014 年 5 月

前言 FOREWORD

　　英语是世界上使用最广泛的语言,也是国际不同母语人群使用最多的沟通媒介,在国际文化和科技交流中起着非常重要的作用。然而,熟练掌握一种语言并非易事,特别是在没有好的语言环境下,要想学好一门语言可能要耗费大量的时间和精力。渗透式双语教学是指在非语言类课程教学过程中以母语为主,穿插使用外语词汇或者用外语对某个名词概念进行讲解的一种教学模式。这种教学模式有两个显著特点,一是可以根据学生对外语掌握的程度,以适当难度将外语融入教学过程,对教师和学生的外语水平没有硬性的要求。二是此种教学过程以较高频度的语码转换模式解释相关知识点的名词、概念以及定理等,使学生不用刻意去背记外语单词即可获得深刻的印象。如果有多门课程采用渗透式双语教学,那么无形中的积累必定会有效地提高学生的外语水平。

　　本书为推动任务型语码转换式双语教学模式在非英语专业中的应用和实践而编写,适用于自动化、测控以及相关的电气信息类专业的渗透式双语教学辅助教材。本书包含基础篇和应用篇两部分。基础篇以专业理论为主,共二十二章,内容涵盖自动化和测控专业基础课和主要专业课,包括传感器与检测技术、电力电子技术、电路原理、计算机控制技术等22门课程的名词及概念。书中所涉及的专业词汇选自近几年出版的优秀中文专业教材,并参考了国外知名大学的原版教材,注重与学科相关的新词、新语。考虑到双语课程需求及专业特点,书中不同程度地渗透了各章节的重要概念、定义和章节概述或体现章节内容主旨的语句及语段。通过专业词汇的中英文对照和典型的中英文语句及段落对照编排,为教师和学生提供相关专业课程的双语教学内容,以达到提高学生专业英语运用能力的目的。应用篇主要是与专业相关的一些应用短文,共三章,包含课程简介、实验报告、实践项目简介以及与专业内容相关的面试用语和实验设备、常用软件的介绍等。编排此部分的目的是为学生提供一些简单的应用模板,借助对这些模板的阅读和理解,实现提高学生英文撰写和英文表达能力的目的。

　　基础篇编写人员有张秀峰(第一章)、刘岩川(第二章)、付立军(第三章)、陈晓云(第四、十六、十八章)、张艳(第五、七章)、王娟(第六章)、韩志敏(第八、十章)、孙进生(第九、十一章)、谢春利(第十二章)、赵凤强(第十三、十四、十九章)、孙炎辉(第十五章)、郭金来(第十七章)、杜海英和徐国凯(第二十章)、薛原(第二十一章)、于为民(第二十二章);应用篇编写人员有谢春利(第二十三章第一节)、张艳(第二十三章第二节)、刘长红(第二十四章)、陈晓云(第二十五章第一节)、杜海英(第二十五章第二节)、刘岩川(第二十五章第三节)、付立军(第二十五章第四节)。曹琳参与了部分章节内容的撰写和修改工作。全书由刘岩川和张艳统稿,张艳为本书的修改和编排做了大量工作。

本书可作为自动化和测控本科专业以及相关专业学生的双语教学辅助用书,也可供电气信息类工程技术人员参考使用。

由于编写时间仓促以及编者水平有限,书中错误和疏漏之处在所难免,敬请广大读者批评指正。

编 者

2014 年 5 月

使用说明

　　本书主要是面向电气信息类专业的教师和学生编写的,适用于任务型语码转换式双语教学。本书内容以自动化和测控两个专业的技术基础课及主要专业课为背景,共选择了22门课程,基础篇中每章对应一门课程,每小节对应一个知识单元。在每小节中包含了与所在知识单元相关的英语单词和短语,每章的最后还安排一定量的短句和语段。

　　本书除了可以利用课程名和知识单元查找相关英文单词以外,在书的最后编排了索引,以方便读者按照汉语查找对应的英文单词。索引按照汉语拼音的顺序排列相关的汉语词汇,并在汉语词汇后面给出了相应英语单词及所在的章节号。

目录 CONTENTS

>> 第一部分　基础篇 /1

>> 第一章　传感器与检测技术 /1
第一节　传感器概述 / 1
第二节　应力传感器 / 2
第三节　热敏传感器 / 3
第四节　固态传感器 / 3
第五节　光电传感器 / 4
第六节　数字传感器 / 5
第七节　检测技术 / 6

>> 第二章　单片机原理及接口技术 / 7
第一节　微型计算机基础 / 7
第二节　51系列单片机硬件详述 / 8
第三节　指令系统与程序设计 / 10
第四节　中断系统 / 10
第五节　定时器与计数器 / 11
第六节　串行通信接口 / 11
第七节　键盘及显示接口 / 12
第八节　模拟接口技术 / 13

>> 第三章　电机与拖动技术基础 / 14
第一节　绪论 / 14
第二节　电力拖动系统动力学 / 15
第三节　直流电机原理 / 15
第四节　他励直流电动机的运行 / 17
第五节　变压器 / 18
第六节　交流电机电枢绕组的电动势与磁通势 / 19
第七节　异步电动机原理 / 20
第八节　三相异步电动机的启动与制动 / 21
第九节　同步电动机 / 22
第十节　微控电机 / 23

>> 第四章　电力电子技术 / 25
第一节　电力电子器件 / 25
第二节　整流电路 / 27
第三节　直流斩波电路 / 28
第四节　交流—交流变流电路 / 29
第五节　逆变电路 / 30
第六节　PWM控制技术 / 31
第七节　软开关技术 / 31
第八节　组合变流电路 / 32

>> 第五章　电路原理 / 34
第一节　电路模型和电路定律 / 34
第二节　电阻电路的等效变换 / 35
第三节　电阻电路的一般分析 / 36
第四节　电路定理 / 38
第五节　含有运算放大器的电阻电路 / 39
第六节　储能元件 / 40
第七节　一阶电路和二阶电路的时域分析 / 41
第八节　相量法 / 43
第九节　正弦稳态电路的分析 / 44
第十节　含有耦合电感的电路 / 46
第十一节　电路的频率响应 / 47
第十二节　三相电路 / 48
第十三节　拉普拉斯变换 / 49
第十四节　二端口网络 / 50

>> 第六章　电气控制技术与PLC应用 / 51
第一节　电气控制技术 / 51
第二节　可编程控制器概述 / 51
第三节　组件与系统 / 52
第四节　PLC编程 / 53
第五节　顺序控制系统 / 54
第六节　通信与网络 / 55
第七节　闭环控制系统 / 55

>> 第七章　电气设计CAD软件 / 57
第一节　项目 / 57
第二节　导线 / 57
第三节　原理图 / 58
第四节　面板布局 / 58
第五节　生成报告 / 59

>> 第八章　工控组态软件 / 60

>> 第九章　工业数据通信与网络技术 / 62
第一节　概论 / 62
第二节　基本原理 / 62
第三节　串行通信标准 / 63
第四节　差错检测 / 64
第五节　开放系统互连模型 / 64
第六节　工业数据通信协议 / 65
第七节　局域网 / 65

>> 第十章 过程控制工程 / 67
第一节 绪论 / 67
第二节 过程建模和过程检测控制仪表 / 67
第三节 单回路控制系统的工程设计 / 68
第四节 复杂过程控制系统 / 69

>> 第十一章 集散控制系统与现场总线技术 / 72
第一节 概述 / 72
第二节 DCS 的结构、硬件和通信 / 72
第三节 软件设计 / 73
第四节 现场总线技术 / 74

>> 第十二章 计算机控制技术 / 76
第一节 绪论 / 76
第二节 模拟量输入输出通道接口技术 / 77
第三节 人机接口技术 / 78
第四节 通用的控制程序设计 / 78
第五节 总线接口技术 / 79
第六节 过程控制的数据处理 / 80
第七节 数字 PID 算法 / 81
第八节 直接数字控制算法 / 82
第九节 模糊控制技术 / 83
第十节 微型计算机控制系统设计 / 84

>> 第十三章 精密机械与仪器 / 85
第一节 绪论 / 85
第二节 机械工程常用材料及钢的热处理 / 85
第三节 平面机构的结构分析 / 86
第四节 平面连杆机构 / 86
第五节 凸轮机构 / 87
第六节 齿轮传动 / 88
第七节 带传动 / 89
第八节 轴、联轴器、离合器 / 90
第九节 支承 / 90
第十节 零件的精度设计与互换性 / 92

>> 第十四章 智能楼宇 / 94
第一节 绪论 / 94
第二节 楼宇自动化控制技术基础 / 94
第三节 楼宇设备自动化系统 / 95
第四节 火灾自动报警与控制 / 96
第五节 楼宇安全防范技术 / 96
第六节 综合布线技术 / 97

>> 第十五章 无线传感器系统 / 99
第一节 概述 / 99
第二节 初识 ZigBee / 99
第三节 ZigBee 应用 / 100
第四节 ZigBee,ZDO 和 ZDP / 101
第五节 网络层 / 103
第六节 ZigBee 开发环境 / 105

>> 第十六章 误差理论与数据处理 / 106
第一节 绪论 / 106
第二节 误差的基本性质与处理 / 107
第三节 误差的合成与分配 / 108
第四节 测量不确定度 / 110
第五节 线性参数的最小二乘法处理 / 110
第六节 回归分析 / 111

>> 第十七章 印刷电路板设计 / 113
第一节 概述 / 113
第二节 原理图设计 / 114
第三节 原理图编辑 / 115
第四节 原理图库文件编辑 / 116
第五节 PCB 设计 / 116
第六节 PCB 设计规则设置 / 117
第七节 PCB 元件封装库编辑 / 118

>> 第十八章 运动控制系统 / 119
第一节 闭环控制的直流调速系统 / 119
第二节 双环控制的直流调速系统 / 120
第三节 可逆调速系统 / 121
第四节 脉宽调制的直流调速系统 / 122
第五节 交流调压调速系统 / 123
第六节 异步电动机变压变频调速系统
——转差功率不变型的调速系统 / 124
第七节 绕线转子异步电动机串级调速系统
——转差功率回馈型的调速系统 / 125

>> 第十九章 智能仪器设计 / 127
第一节 概述 / 127
第二节 数据采集技术 / 127
第三节 人机对话与数据通信 / 128
第四节 智能仪器的基本数据处理算法 / 129
第五节 软件设计 / 130
第六节 可靠性与抗干扰技术 / 130
第七节 可测试性设计 / 131

目录 CONTENTS

>> 第二十章 自动控制理论 / 133
第一节 绪论 / 133
第二节 控制系统的数学模型 / 133
第三节 控制系统的时域分析 / 134
第四节 根轨迹 / 135
第五节 频率响应法 / 135
第六节 控制系统的校正 / 136
第七节 PID控制与鲁棒控制 / 137
第八节 离散控制系统 / 137
第九节 状态空间分析法 / 138
第十节 非线性控制系统 / 139

>> 第二十一章 数字电子技术 / 141
第一节 绪论 / 141
第二节 逻辑代数基础 / 141
第三节 逻辑门电路 / 142
第四节 基础组合逻辑电路 / 143
第五节 锁存器和触发器 / 144
第六节 时序逻辑电路 / 144
第七节 脉冲波形的产生与整形 / 145
第八节 半导体存储器 / 146
第九节 可编程逻辑器件 / 146
第十节 数模与模数转换器 / 147

>> 第二十二章 模拟电子技术 / 148
第一节 绪论 / 148
第二节 运算放大器 / 148
第三节 二极管及其基本电路 / 149
第四节 双极结型三极管及放大电路基础 / 150
第五节 场效应管放大电路 / 151
第六节 模拟集成电路 / 151
第七节 放大电路中的反馈 / 152
第八节 功率放大电路 / 152
第九节 信号的处理与信号的产生电路 / 153
第十节 直流电路 / 154

>> 第二部分 应用篇 / 155

>> 第二十三章 应用范例 / 155
第一节 摘要写作 / 155
第二节 实验报告 / 157

>> 第二十四章 设备使用手册 / 161

>> 第二十五章 简介与对话 / 172
第一节 《电力电子技术》课程简介 / 172
第二节 矩阵实验室(MATLAB)简介 / 172
第三节 实践项目简介 / 174
第四节 就业情景对话 / 176

>> 参考文献 / 179

>> 索 引 / 181

第一部分 基础篇

第一章 传感器与检测技术
Chapter 1 Sensor and Measure Technology

第一节 传感器概述
Section 1 Introduction of Sensor

测量电路　measuring circuit
迟滞　hysteresis
传递函数　transfer function
传感器　sensor/ transducer
粗大误差　gross error
动态特性　dynamic characteristic
非电量检测仪　measuring instrument for non-electric quantity
非线性误差　non-linearing error
分辨率　resolution
检测技术　detection technique
精度　precision
精确度　accuracy
静态特性　static characteristic
灵敏度　sensitivity
零点漂移　zero drift

脉冲　impulse
敏感元件　sensing element
偏差　excursion
随机误差　random error
温漂　temperature drift
物理量检测仪表　measuring instrument for physical quantity
误差　error
系统误差　system error
线性化　linearization
校正　remedy
仪表　gauge / meter
智能仪器　intelligent instrument
重复性　repeatability
转换元件　switch element
最小二乘法　least square method

[1] The sensors are transforming the analog physical signal into a computing platform. Modern sensors not only respond to physical signals to produce data, they also embed computing and communication capabilities. They are thus able to store, process locally and transfer the data they produce over a long distance.

传感器可把模拟的物理量转换为可测的数值量。随着现代技术的发展,传感器不仅可以转换模拟信号,而且还具有计算和通信的功能。因此,现代传感器可以在本地储存、处理数据,并能够远距离传送采集到的数据。

[2] Detection technology is an important means of information acquisition. A variety of non-electric quantity information could be measured by the appropriate method and apparatus using a variety of physical and chemical effects. Detection technology is a necessary component in the field of automation, and it is the premise and basis of automatic control.

检测技术是获取信息的重要手段,利用各种物理、化学效应,通过合适的方法与装置,实现各种非电量信息的测量。检测技术是自动化领域的重要组成部分,是自动控制的前提和基础。

第二节　应力传感器
Section 2　Strain Pressure Sensor

半导体应变式压力传感器　semiconductor strain pressure sensor
边缘效应　contour effect
变换特性　changing characteristic
不平衡电容　unbalanced capacitance
差动变压器　differential transformer
差动电容传感器　differential capacitance sensor
差压变送器　differential pressure transmitter
差压传感器　differential pressure pick-up
差压指示器　different pressure indicator
等臂电桥　equal-arm bridge
等效电路　equivalent electric circuit
电动测微仪　electronic gage
电感线圈　induction coil
电感元件　induction element
电极栅　electrode grid
电流互感器　current transformer
电桥　bridge
电容式传感器　capacitive sensor
电容式压力变送器　capacitive pressure transmitter
电容式液位探头　capacitive liquid-level probe
电涡流式传感器　electric swirl sensor
电压放大器　voltage amplifier
调幅　amplitude modulation
调频　frequency regulation
动态响应　dynamic response
二极管检波电路　diode wave detector
分布电容　distribution capacity
横向效应　transverse effect
互感式传感器　mutual sensor
激励电压　energizing voltage
寄生电容　parasitic capacitance
加速度传感器　acceleration sensor
力敏元件　mechanical sensor
临界压力　critical pressure
零点残余电压　zero residual voltage

气隙　air gap
石英传感器　quartz sensor
特性分析　characteristic analysis
位移测量　displacement measure
位置变送器　position transmitter
温度补偿　temperature compensation
温度误差　temperature error
无源元件　passive component
误差因数　error factor
谐振曲线　resonant curve
压电传感器　piezoelectric transducer
压电式加速度计　piezoelectric accelerometer
压电式流量计　piezoelectric flow meter
压电式压力传感器　piezoelectric pressure sensor
压电陶瓷　piezoelectric ceramic
压电效应　piezoelectric effect
压力传感器　pressure pick-up / pressure transducer
压敏电桥　voltage-sensitive bridge
压敏电阻　voltage dependent resistor
压敏二极管　pressure-sensitive diode
压阻式传感器　compressive resistance sensor
压阻系数　compressive resistance coefficient
压阻效应　compressive resistance effect
应变极限　strain limitation
应变片　strain gauge/ foil gauge
应变式压力传感器　strain pressure sensor
应力传感器　strain sensor
有源滤波器　active filter
运算放大器　operation amplifier
振幅测量　vibration amplitude measure
整流电路　rectification circuit
智能压力变送器　intelligent pressure transmitter
转换测量　conversion measure
自感式传感器　self-induction sensor
阻抗变换器　impedance transformer

1 Strain sensor could transform the non-electric quantity measured into the change of the circuit element parameters, using of the effect of strain and elastic deformation of components, so the physical quantity be measured. According to the difference of parameters, stress sensor is divided into resistance, capacitance and inductance sensor. They can be used to measure the deformation, pressure, stress, displacement, acceleration, vibration, flow etc. as non electrical quantities.

应力传感器是利用元件的应变效应或弹性形变，将被测非电量转换成电路元件参数的变化实现测量的装置，根据元件参数的不同，应力传感器可分为电阻式、电容式和电感式传感器，可用于测量形变、压力、力、位移、加速度、振动、流量等非电量。

2 Compared with optical, inductive and piezoresistive transducers, capacitive sensors have the following advantages: low cost and power usage, good sensitivity, and speed. Capacitive sensors can detect motion, acceleration, flow, and many other variables, and are used in a wide range of applications.
相对于光电传感器、感应传感器以及压阻式传感器而言,电容式传感器有如下一些优点:价格低、耗电省、灵敏度高、速度快。电容式传感器能够检测运动、加速度、流量以及很多其他变量,因此获得了广泛的应用。

第三节 热敏传感器
Section 3 Thermo-electric Sensor

NFF型温度传感器　NFF temperature sensor
标称电阻　standard resistance
铂电阻　platinum resistance
测温电路　temperature measure circuit
电阻—温度特性　electric resistance-temperature characteristic
伏安特性　current-voltage characteristic
干涉式温度传感器　disconcerting temperature sensor
红外线式温度传感器　temperature sensor through infrared ray
环境温度　ambient temperature
集成温度传感器　integrated temperature sensor
继电效应　relay effect
接触电势　contact potential
冷端补偿器　cold junction compensator
冷端温度　cold end temperature
热磁效应　thermo-magnetic effect

热电极　hot electrode
热电偶　thermocouple
热电桥　thermal bridge
热电效应　thermo-electric effect
热电阻　thermal resistance
热敏电阻器　thermal resistor
热敏开关　thermoswitch
热敏元件传感器　thermistor element sensor
铜电阻　copper resistance
温差电流　thermocurrent
温度传感器　temperature sensor
温度系数　temperature coefficient
无稳态多谐振荡器　astable multivibrator
荧光辐射式温度传感器　temperature sensor through fluorescent radiation
遮光路式温度传感器　hide temperature sensor

1 Thermoelectric sensor is the device that could transform the change of temperature into electric quantity. The thermoelectric sensors transform the change of temperature into the change of voltage or resistance, and can be divided into thermocouple, thermal resistance and thermistor.
热电式传感器是一种将温度变化转换为电量实现测量的元件,热电式传感器可以将温度的变化转换为电势或电阻的变化,分为热电偶、热电阻和热敏电阻。

2 A thermistor is a temperature sensor that has been developed by using semiconductor material. The principle applied is that semiconductor resistance changes with temperature. Particular semiconductor material used varies widely to accommodate temperature ranges, sensitivity, resistance ranges and other factors.
热敏电阻是一种用半导体材料制成的温度传感器。它所利用的原理是当周围温度改变时,半导体材料的电阻也随之改变。为了满足测量温度时要求的测量范围、灵敏度、电阻范围以及其他测量要求,往往采用一些特殊的半导体材料。

第四节 固态传感器
Section 4 Solid State Sensor

半导体气体传感器　semiconductor gas sensor
磁电特性　magnetoelectric characteristic
磁敏传感器　magneto sensor
磁敏二极管　magneto diode

磁敏三极管　magneto triode
磁致伸缩式传感器　magnetostrictive sensor
磁阻传感器　magnetic resistance sensor
电磁效应　electromagnetic effect

自动化与检测技术

电光效应	electric optical effect
电液型	electro hydromatic type
法拉第效应式传感器	Faraday's effect-sensor
感应电势	inducer electromotive force
固态传感器	solid-state sensor
固态图像传感器	solid-state image sensor
固态温度传感器	solid-state temperature sensor
光电池	photo battery
光电传感器	photoelectric sensor
光电导探测器	photoconductive detector
光电效应	photoelectric effect
光电元件	photoelectric unit
光电转速计	photoelectric speed recorder
光敏传感器	light sensitive sensor
光敏电阻	light sensitive resistance
光敏二极管	light sensitive diode
光敏区	light sensitive zone
光谱特性	light spectrum characteristic
霍尔系数	Hall coefficient
霍尔效应	Hall effect
霍尔压力变送器	Hall pressure transmitter
霍尔元件	Hall element
寄生直流电势	parasitic direct voltage
绝对湿度	absolute humidity
力敏效应	mechanical effect
流量变送器	flow transducer
光敏三极管	light sensitive triode
气动量仪	air gage
气敏元件	gas sensor
数字式电液控制系统	digital electro-hydromatic control
陶瓷温度传感器	ceramic temperature sensor
温度补偿电路	temperature compensation circuit
稳定性	stability
涡流传感器	eddy sensor
线形固态图像传感器	linear solid state image sensor
相对湿度	relative humidity
行传输	line transmission
液位变送器	level transducer
帧传输	frame transmission
转速传感器	rate sensor / speed sensor

1 Solid-state sensor measures the non-electrolyte quantity using the change of physical properties when the materials are effected by the external factors. The sensitive materials of sensor are mainly semiconductor, electrolyte and ferroelectrics. And this kind of sensor has the advantages of simple structure, small volume, good dynamic response, the output of electric quantity, low power consumption, safety and reliability, etc.

固态传感器是指利用材料在外部因素作用下物理性质发生变化而实现非电量测量的传感器。这类传感器主要以半导体、电介质、铁电体等为敏感材料，具有结构简单、体积小、动态响应好、输出为电量、低功耗、安全可靠等特点。

2 Integrated circuit technology has led to the development of solid-state (SS) pressure sensors that find extensive application in the pressure ranges of 0 to 100 kPa. These small units often require no more than three connections——DC power, ground, and the sensor output. Pressure connection is via a metal tube.

随着集成电路技术的发展，固态压力传感器也得到了极大的发展。目前，固态压力传感器主要用于测量压力值为 0 到 100 千帕的场合。固态压力传感器体积一般都很小，而且只有电源端、接地端和传感器输出端。压力连接是通过一根金属管。

第五节 光电传感器
Section 5 Photoelectric Sensor

CCD 图像传感器	CCD image sensor
半导体激光器	semiconductor energizer
标定	calibrate
标定量程	calibrate span
动态标定	dynamic calibrate
共振式光导纤维振动传感器	resonant optical fiber vibration sensor
固体激光器	solid energizer
光导纤维	optical fiber
光导纤维 FF 型传感器	optical fiber FF sensor
光电检测器	photodetector
光电脉冲发生器	optoelectronic pulse generator
光纤传感器	optical fiber sensor
光纤磁场电流传感器	optical fiber sensor for

	magnetic field and electric current
光纤水声传感器　optical fiber sensor for water sound	静态标定　static calibrate
光纤转速传感器　optical fiber sensor for turn speed	偏振调制　polarization modulation
光源　light source	气体激光器　gas energizer
激光扫描测量装置　laser scanner	强度调制　strength modulation
集成红外CCD固态图像传感器　integrated	相位调制　phase modulation
	infrared CCD solid-state image sensor

1 Photoelectric sensor is used to measure the optical signal or measure the other physical quantities that have been transformed into optical signal, and the outputs of sensor is the electrical signal. The kind of sensor consists of the light path and the circuit. In according to the optical theory, there are four types of photoelectric sensors, which are transmissive, reflective, radiation and switch type.

光电式传感器可用来测量光信号或测量已经转换为光信号的其他物理量,然后输出电信号。光电式传感器由光路及电路两大部分组成,根据光学原理可以分为透射式、反射式、辐射式和开关式四种类型。

2 The photodiode effect refers to the fact that photons impinging on the PN junction also alter the reverse current-verse-voltage characteristic of the diode. In particular, the reverse current will be increased almost linearly with light intensity. Thus, the photodiode is operated in the reverse-bias mode.

光敏二极管效应指的是PN结附近受到光子轰击,二极管的反向电流随电压变化而变化。也就是说,随着光照强度的增加,反向电流也随之增加,变化趋势呈线性。因此,光敏二极管以反向偏置的形式工作。

第六节　数字传感器
Section 6　Digital Sensor

磁栅传感器　magnetic grid sensor	脉冲式数字传感器　impulse digital sensor
电磁法　electromagnetic method	模数转换器　analog-to-digital converter
二进制码　binary-code	位移传感器　displacement sensor
感应同步器　inductosyn	位置变送器　position transmitter
感应同步式　induction synchronous	循环码　cyclic code
固有频率　free running frequency	振簧式频率传感器　vibrating-reed frequency sensor
光栅传感器　raster sensor	
码盘　coded disc	

1 Digital sensor could transform the non-electrical quantity into digital signal directly. Digital sensor has many characteristics, such as high measuring accuracy and resolution, strong anti-interference ability, and convenient for computer interface, suitable for long-distance transmission. The output of the sensor has mainly two forms——coding mode and output counting.

数字式传感器是能够直接将非电量转换为数字量的传感器,它具有测量精度和分辨率高、抗干扰能力强、便于与计算机接口、适合于远距离传输的特点,主要有编码方式和输出计数两种输出形式。

2 One major issue with a mobile robot acting as a gateway is the communication between the robot and the sensor network. Sensor networks typically communicate by using 900 MHz radio waves. Mobile robots use laptops that communicate via 802.11, in the 2.4 to 2.483 GHz range. Intel hopes to prove that a sensor net can be equipped with 802.11 capabilities to bridge the gap between robotics and wireless networks.

移动机器人装置能够起关键作用的一个问题是机器人装置与传感器网络的通讯问题。通常情况下传感器网络使用900 MHz的无线电波进行通讯。移动机器人使用膝上型电脑通讯,经过802.11协议,也就是在 2.4 到 2.483GHz 的范围内。英特尔公司希望能生产一种传感器,具有 802.11 协议需要的能力,在机器人系统和无线网络间提供通讯。

第七节　检测技术
Section 7　Measure Technology

波纹管　bellow pipe
差压　differential pressure
差压式流量计　differential pressure flowmeter
成分检测　component detection
弹簧管　spring pipe
弹性元件　elastic component
电磁流量计　electromagnetic flowmeter
非接触式测温　non-contact temperature measurement
厚度测量　thickness measurement
机械量检测　mechanical quantity detection
接触式测温　contact temperature measurement
力学量测量　dynamics quantity measurement
流量计　flow meter

流量检测　flow detection
膜盒　capsule
体积流量　volume flow
位移测量　displacement measurement
温标　temperature scale
温度变送器　temperature transmitter
温度检测　temperature detection
涡街流量计　vortex flowmeter
物位检测　material level detection
压力检测　pressure detection
质量流量　mass flow
转速测量　speed measurement

Detection technology is an important part of automation subject. Along with the rapid development of science, technology and with the appearance of new materials, detection technology has changed fundamentally in the basic theories, system mechanism, design procedures and experimental methods. The factors that drive detection technology improving are micro-electronics, computer and communication technologies. The detection technology is going towards digitalization, networking and intelligent ualization direction.

　　检测技术是自动化学科的重要组成部分。随着科学技术的迅速发展以及新材料出现,检测技术这个学科在基础理论、系统机构、设计程序和实验方法上都发生了根本性的变革。驱动检测技术进步的因素包括微电子、计算机和通信技术。检测技术正向着数字化、网络化和智能化的方向发展。

第二章 单片机原理及接口技术
Chapter 2　Single Chip Microprocessor Principles and Interface

第一节　微型计算机基础
Section 1　Foundation of Microcomputer

16 位机　16-bit computer
4 位机　4-bit computer
8 位机　8-bit computer
ASCII 码　American Standard Code for Information Interchange
BCD 码　binary code decimal
EPROM 编程器　EPROM programmer
八进制　octal system
补码　base complement
不挥发存储器　nonvolatile memory
乘法器　multiplier
除法器　divider
存储　store
存储器　memory
打印机　printer
打印头　print head
大型计算机　mainframe
单片机　single-chip microcomputer
地址缓冲器　address buffer
动态存储器　dynamic RAM
二进制　binary system
反码　base minus one's complement
访问　access
浮点数　floating-point data
个人计算机　personal computer
工作寄存器　working register
光盘　compact disc
光驱　CD driver
缓冲器　buffer
换行　line feed
回车　carriage return
绘图仪　graph plotter
基数　radix
集成电路　integrated circuit
寄存器　register
加法器　summator
减法器　subtracter

阶码　exponent
静态存储器　static RAM
可擦写的只读存储器　erasable programmable read-only memory(EPROM)
控制器　controller
累加器　accumulator
目的　destination
内存条　memory bank
偏移量　offset
取址　take address
软件　software
十进制　decimal system
十六进制　hexadecimal system
实数　real data
鼠标　mouse
数据缓冲器　data buffer
双字　double word
随机存取存储器　random access memory
微处理器　microprocessor
微型计算机　microcomputer
尾码　mantissa
位　bit
显示器　display
小数/分数　fraction
小型计算机　minicomputer
音箱　sound box
硬件　hardware
硬盘　hard disk
原码　original code
源　source
整数　integer
执行　execute
只读存储器　read only memory
中央处理器　central processing unit(CPU)
紫外线　ultraviolet light
字　word
字节　byte

1 In the past few years, great progress has been made in the cost, function, and the use of microcomputers. It is now possible, with a relatively small expenditure, to purchase a microcomputer system that will take data, quickly analyze them, and show the results or control a process. This has been made possible by the development of technology that can fabricate millions of transistors, diodes, resistors, capacitors, and conductors on a single silicon integrated circuit chip. The 8086 is a 16-bit microprocessor based on the 8080. The 8086 uses silicon-gate HMOS technology for faster performance and an expanded 8080 structure. Basically, it is an improved 16-bit version of 8080. The 8080 multiplexed bus has been expanded into a 16-bit external bus. Like the 8080, the instructions are byte-oriented.

在过去的几年中，微型计算机在其控制成本、提高性能和使用方面取得了很大的进步。现在，我们花少量的费用就可以购买一个微型计算机系统来完成数据的收集和快速分析并显示结果或者控制一个过程。这完全仰仗于可将成百万个晶体管、二极管、电阻、电容以及导体制作在一块硅集成电路芯片上的技术。8086是建立在8080基础上的十六位微处理器。为了提高工作速度，扩展8080结构，8086利用了硅栅金属氧化物半导体技术。在本质上讲，8086是改进的8080的十六位版本。8080的多路总线已扩展成十六位外部总线。和8080一样，其指令也是面向字节的。

2 Microprocessor performs three main tasks for the computer system: (1) data transfer between itself and memory or I/O systems, (2) simple arithmetic and logic operations and (3) program flow via simple decision.

微处理器在计算机系统中的三个主要作用是:(1)微处理器与存储器或I/O系统之间的数据传输，(2)简单的算术和逻辑运算，(3)通过简单的判断决定程序的流程。

第二节 51系列单片机硬件详述

Section 2 51 Series Single Chip Microprocessor Hardware Specification

中文	英文	中文	英文
TTL兼容	TTL compatible	发光二极管	light-emitting diode
编码器	encoder	翻转电路	flip-flop circuit
编址	addressing	反向器	reviser
波形	waveform	访问时间	access time
程序存储器扩展	program memory expending	辅助进位标志	auxiliary carry flag
程序区	program area	负载	load
初始化	initialization	负载能力	load capacity
触发器	trigger	复位	reset
磁盘驱动器	disk driver	复用	multiplexed
存储器读信号	memory reading signal	高电平触发	high level trigging
存储器扩展	memory expanding	高阻态	high impedance
存储器容量	memory capacity	缓冲	buffer
存储器写信号	memory writing signal	机器周期	machine cycle
低电平触发	low level trigging	集成	integration
地址锁存	address latching	检测	detect
地址锁存器	address latch	键盘	keyboard
地址锁存信号	address latching signal	接口	interface
地址信号	address signal	接口电路	interface circuit
地址总线	address bus	进位标志	carry flag
掉电	power down	晶振	crystal oscillator
独立I/O	isolated I/O	晶振频率	crystal oscillator frequency
端口地址	port address	可编程接口	programmable interface
堆栈	stack	控制信号	control signal

第二章 单片机原理及接口技术

控制总线　control bus
临时　temporary
流程图　flowchart
逻辑电平　logic level
脉冲　pulse
模拟信号　analog signal
模数转换　analog to digital conversion
内部存储器　internal memory
内部地址总线　internal address bus
内部控制总线　internal control bus
内部数据总线　internal data bus
内部总线　internal bus
片选　chip selection
片选信号　chip selection signal
奇偶　parity
驱动　drive
软件　software
上拉电阻　pull-up resistor
时序　timing
时钟周期　clock period
史密特触发器　Schmitt trigger
输出缓冲器　output buffer
输出允许　output enable
输入缓冲器　input buffer
输入输出端口　I/O port
数据存储器扩展　data memory expending
数据缓冲器　data buffer
数据区　data area
数据锁存信号　data latching signal
数据信号　data signal
数据总线　data bus
数模转换　digital to analog conversion

数字信号　digital signal
双向的　bi-directional
双向数据线　bi-directional data wire
锁存　latching
锁存器　latch (unit)
特殊功能寄存器　special function register(SFR)
外部存储器　external memory
外部地址总线　external address bus
外部控制总线　external control bus
外部数据总线　external data bus
外围的　peripheral
握手信号　handshaking signal
下拉电阻　pull-down resistor
显示　display
协处理器　coprocessor
写允许　write enable
选通　strobe
选通信号　strobe signal
阳极　anode
一兆(百万)　mega
译码法编址　decoder addressing
译码器　decoder
溢出　overflow
溢出标志　overflow flag
阴极　cathode
应答　acknowledge
元件　component
指令周期　instruction cycle
置位　set
总线　bus
总线周期　bus cycle

1 All computer systems have three buses：(1)an address bus that provides the memory and I/O port number，(2)a data bus that transfers data between the microprocessor and memory and I/O system, and (3)a control bus that provides control signals to the memory and I/O.
所有的计算机系统都有三种总线：(1)地址总线用以提供存储器和 I/O 端口的地址，(2)数据总线用于在微处理器和存储器及 I/O 系统之间传输数据，(3)控制总线为存储器和 I/O 提供控制信号。

2 The Serial Peripheral Interface (SPI) is an industry standard synchronous serial interface that allows eight bits of data to be synchronously transmitted and received simultaneously. The system can be configured for master or slave operation.
串行外围接口(SPI)是工业标准的同步串行接口，它允许八位数据同时同步地被发送和接收。系统可设置为主操作或从操作模式。

3 The 8051 series of microcontrollers are highly integrated single chip microcomputers with an 8-bit CPU, memory, interrupt controller, timers, serial I/O and digital I/O on a single piece of silicon.
8051 系列单片机是高度集成的微处理器，在其硅片上有八位 CPU、存储器、中断控制器、定时器及串行和数字接口。

4 The five memory spaces of the 8051 are: program memory, external data memory, internal data memory, special function registers and bit memory.
8051 的五类存储器分别是程序存储器、外部数据存储器、内部数据存储器、特殊功能寄存器和位存储器。

第三节　指令系统与程序设计
Section 3　Instruction Set and Program Design

保留符号　reserved symbol	逻辑运算　logical calculus
变量　variable	目的操作数　destination operand
变量表　variable table	起始地址　beginning address
布尔运算　Boolean operation	算术运算　arithmetic operation
操作码　opcode	位存储器地址　bit memory address
操作数　operand	位寻址　bit addressing
常数　constant	无符号数　unsigned number
常数表　constant table	无符号数加法　unsigned number addition
程序结束　program end	无符号数减法　unsigned number subtraction
程序名　program name	寻址方式　addressing mode
错误信息　error message	循环初态　loop initial state
汇编　assembling	循环结束　loop end
汇编程序　assembly program	循环结束条件　loop end condition
汇编错误码　assembler error code	循环体　loop body
汇编符号集　assembler character set	有符号数　signed number
汇编语言　assembly language	源操作数　source operand
机器语言　machine language	直接寻址　direct addressing
基址加变址寻址　base plus index addressing	指令　instruction
寄存器间接寻址　register indirect addressing	助记符　mnemonic
寄存器寻址　register addressing	字符串　character string
控制转移　control transfer	最低位　low order position
立即寻址　immediate addressing	最高位　high order position

Efficient software development for the microprocessor requires a complete familiarity with the addressing modes used by each instruction. The data-addressing modes include immediate, direct, register-indirect and base-plus-index, in the 8051 single chip microprocessor.

微处理器的高效的软件开发需要对每种指令所采用的寻址方式十分熟悉。8051 单片机的数据寻址方式包括立即寻址、直接寻址、寄存器间接寻址、基址加变址寻址。

第四节　中断系统
Section 4　Interrupt System

保护现场　protecting locality	禁止中断　disable interrupt
边沿触发　edge triggering	控制位　control bit
触发　triggering	内部中断　internal interrupt
串行口中断　serial port interrupt	入口地址　entry address
电平触发　level triggering	外部中断　external interrupt
定时器中断　timer interrupt	溢出中断　overflow interrupt
复位状态　reset state	优先级　priority level
恢复现场　resuming locality	中断　interrupt
计数器中断　counter interrupt	中断返回　return from interruption

中断服务程序　interrupt service routine
中断请求　interrupt request
中断请求标志　interrupt request flag
中断入口地址　interrupt entry address
中断条件　interrupt condition
中断响应　interrupt response
中断源　interrupt source
中断允许　interrupt enable

The IP register sets one of the two main priority levels for the various interrupt sources. Set the corresponding bit to "1" to configure interrupt as high priority, and to "0" to configure interrupt as low priority.

寄存器 IP 为各个中断源设置两个主要的中断优先级。把相应的位置设置为"1"，可以把对应中断源的优先级设置为高优先级；设置为"0"，则把该中断源的优先级设置为低优先级。

第五节　定时器与计数器
Section 5　Timer and Counter

倍频　frequency doubling
边沿　porch
查询　enquiry
查询方式　enquiry mode
初始值　initial value
等效逻辑电路　equivalent logic circuit
电平　electrical level
定时器　timer
定时器方式　timing mode
方波　square wave
方式选择位　mode select bit
分频　frequency division
负脉冲　minus pulse
工作方式　operation mode

关中断　disable interrupt
计数长度　count length
计数器　counter
计数器方式　counting mode
开中断　enable interrupt
启动计数　start counting
停止计数　stop counting
外部输入脉冲　external input pulse
运行控制位　executive control bit
振荡频率　oscillation frequency
正脉冲　plus pulse
中断标志位　interrupt flag bit
中断溢出　interrupt overflow

The 8051 has two 16-bit timer/counters, namely: timer 0 and timer 1. Each timer/counter consists of two 8-bit registers THx and TLx (x = 0 and 1). Both can be configured to operate either as 16-bit timers or event counters. In "timer" function, the TLx register is incremented by every machine cycle. Thus one can think of it as counting machine cycles. Since a machine cycle has 12 oscillator periods the maximum count rate is 12 times of the oscillator frequency.

8051 有两个十六位定时/计数器，分别为定时器 0 和定时器 1。每一个定时器/计数器包含两个八位寄存器，即 THx 和 TLx(x=0 和 x=1)。每一个定时/计数器都可以设定成一个十六位的定时器或事件计数器。在定时器功能时，每一个机器周期使 TLx 寄存器增加，因此可以把机器周期看作计数周期。由于一个机器周期包含十二个振荡周期，因此最大计数速率是振荡频率的十二倍。

第六节　串行通信接口
Section 6　Serial Communication Interface

半双工　semi duplex
并行数据　parallel data
并行通讯　parallel communication
波特　Baud
波特率　Baud rate
波特率发生器　Baud rate generator
串行　serial
串行接口　serial interface

串行数据　serial data
串行通讯　serial communication
地址帧　address frame
调制解调器　modem
多机通讯　multimachine communication
发送端　sending terminal
发送缓冲器　sending buffer
发送控制器　sending controller

发送数据　sending data
发送中断标志　transmission interrupt flag
方式选择位　mode select bit
分布式　distributivity
分布式系统　distributivity system
检测　testing
接收端　receiving terminal
接收缓冲器　receiving buffer
接收控制器　receiving controller
接收数据　receiving data
接收中断标志　reception interrupt flag
空闲位　vacancy bit
奇偶校验位　parity bit
起始位　start bit
全双工　full duplex

数据发送位　data transmission bit
数据位　data bit
数据位低位　low position of data bit
数据位高位　high position of data bit
数据帧　data frame
停止位　stop bit
通讯　communication
通讯方式　communication mode
同步通讯　synchronic communication
异步通讯　asynchronous communication
允许接收位　enable receiver bit
帧　frame
帧格式　frame format
主机　host computer

The serial port of 8051 is full duplex. That's to say it can simultaneously transmit and receive data. It is also receive-buffered, meaning it can start receiving a second byte before a previously received byte has been read from the receiving register.

8051的串行口是全双工的，也就是说它可以同时发送和接收数据。它具有接收缓冲功能，表示在接收寄存器读出已经接收到的一个字节之前，可以接收第二个字节。

第七节　键盘及显示接口
Section 7　Keyboard and Display Interfacing

按键　key press
背光　backlight
出错信号　error signal
击键　keystroke
键码　key code
键盘接口　keyboard interface
键盘接口技术　keyboard interface technology
键盘扫描　keyboard scan
键盘事件　keyboard event
键扫描程序　keyboard scanning procedure
七段数码管显示器　7-Segment LED numeric displayer
输出设备　output device

输入设备　input device
输入输出设备　input and output device
显示　display
显示程序　display routine
显示缓冲区　display data buffer
显示接口　display interface
显示器设计技术　display design technology
显示子程序　display subroutine
液晶　liquid crystal
应答信号　answer-back signal
中断方式　interrupt mode
转换技术　changing technology

In normal usage, the keyboard is used to type texts and numbers into a word processor, text editor or other programs. In a modern computer, the interpretation of key presses is generally left to the software. A computer keyboard distinguishes each physical key from every other and reports all key presses to the controlling software. A keyboard is also used to give commands to the operating system of a computer, such as Windows' Control-Alt-Delete combination, which brings up a task window or shuts down the machine. Keyboards are the only way to enter commands on a command-line interface.

键盘一般用于将字符和数字送入字处理器、编辑器或其他程序。当今计算机的键盘能够区别每一个物理键，把每一次按键报告给监控程序，因此对按键的解释一般留给软件来处理。键盘也用于给出操作系统的命令，像Windows的Control-Alt-Delete组合键能够调出任务窗口或者关闭计算机。在命令行界面中，键盘是输入命令的唯一手段。

第八节　模拟接口技术
Section 8　The Technic of Analog Interfacing

T型电阻解码网络　R-2R ladder network	基准电压　reference voltage
采样　sampling	精度　precision
采样保持器　sampling hold unit	连续量　continuous quantity
参考电压　reference voltage	模拟电流信号　analog current singnal
电流输出端　current output port	模拟电压信号　analog voltage singnal
电压输出端　voltage output port	模拟量　analog quantity
电阻网络　resistance network	模拟滤波　analog filtering
多路 A/D 转换器　multi-channel A/D converter	模数转换器　analog to digital converter
反馈电阻　feedback resistance	数模转换器　digital to analog converter
放大器　amplifier	数字量　digital quantity
分辨率　distinguishability / resolution	线性度　linearity
基准电流　reference current	运算放大器　operation amplifier

　　The resolution of the analog to digital converter indicates the number of discrete values it can produce over the range of analog values. The values are usually stored electronically in binary form, so the resolution is usually expressed in bits. In consequence, the number of discrete values available, or "levels", is a power of two. For example, an ADC with a resolution of 8 bits can encode an analog input to one in 256 different levels, since $2^8 = 256$. The values can represent the ranges from 0 to 255 (i.e. unsigned integer) or from -128 to $+127$ (i.e. signed integer), depending on the applications.

　　模数转换器的分辨率是指转换器在对满量程的模拟量进行转换时所产生的数值。由于此值通常以二进制形式存储，所以，分辨率通常用位数来表示。这样，满量程所得到的数值，或称"水平"，是2的整次幂。例如，一个8位分辨率的模数转换器可以对其输入的模拟量进行256个不同水平的编码，因为 $2^8 = 256$。它可以表示从0到255（无符号整数）或从-128到127（有符号整数）范围内的数值，是否带符号取决于具体应用。

第三章 电机与拖动技术基础
Chapter 3　Fundamentals of Electrical Machines and Drives

第一节　绪论
Section 1　Introduction

安培	Ampere	伏	volt
饱和	saturation	功	work
磁场强度	magnetic field intensity	赫兹	Hertz
磁畴	magnetic domain	机电能量转换	electromechanical energy conversion
磁导率	permeability	基尔霍夫电流定律	Kirchhoff's current law(KCL)
磁化曲线	magnetic curve	基尔霍夫电流方程	KCL equation
磁路	magnetic circuit	基尔霍夫电压定律	Kirchhoff's voltage law(KVL)
磁密	flux density	基尔霍夫电压方程	KVL equation
磁通	magnetic flux	交流电流	alternating current
磁通密度	magnetic flux density	矫顽力	coercive force
磁通势	magnetomotive force	欧姆	ohm
磁性材料	magnetic material	欧姆表	ohmmeter
磁滞回线	hysteresis loop	去磁效应	demagnetization effect
磁滞损耗	hysteresis loss	剩磁	magnetic remanence
导体	conductor	铁磁材料	ferro-magnetic material
电磁感应	electromagnetic induction	拖动	drive
电机	electrical machine	瓦	watt
电流	current	涡流	eddy current
电势	electromotive force	西门子	Siemens
电压	voltage	永磁铁	permanent magnet
电压降	voltage drop	匝数	number of turns
法拉	Farad		

1 Faraday's law of electromagnetic induction:
  If the flux linking a loop (or turn) varies as a function of time, a voltage is induced between its terminals; its value is proportional to the rate of change of flux.
  法拉第电磁感应定律:
  如果磁通链着线圈,当磁通变化时,线圈两端感应电动势,其大小与磁通变化率成正比。

2 Transformers and most electric motors operate on alternating current. In such devices the flux in the iron changes continuously both in value and direction. If we plot the flux density B as a function of H, we obtain a closed curve called hysteresis loop.
  变压器和多数电动机运行在交流电流下。在这些设备中,铁芯中的磁通在大小和方向上都连续变化。如果画出磁密 B 随 H 的函数关系,所得到的封闭曲线称作磁滞回线。

3 The significance of remanent magnetization is that it can produce magnetic flux in a magnetic circuit in the absence of external excitation.
  剩磁的意义在于,当没有外部励磁存在时,它也能在磁路中产生磁通。

第二节　电力拖动系统动力学
Section 2　Kinetics of Electrical Driving System

参考方向	reference direction	切削	cut
齿轮	gear	热能	thermal energy
电动机	motor	势能	potential energy
动能	kinetic energy	输出功率	output power
动态平衡	dynamic equilibrium	输入功率	input power
飞轮矩	flywheel torque	位移	displacement
负载	load	温标	temperature scale
功率	power	稳态	steady-state
惯性	inertia	效率	efficiency
过渡过程	transient process	制动	brake
滑轮	pulley	重力	force of gravity
机械能	mechanical work	转动惯量	moment of inertia
加速	acceleration	转矩	torque
减速	deceleration	转速	rotational speed / speed of revolution
切向力	tangential force		

1 When a motor drives a mechanical load, the speed is usually constant. In this state of dynamic equilibrium, the torque T developed by the motor is exactly equal and opposite to the torque T_L imposed by the load. However, if the motor torque is raised so that it exceeds the load torque, the speed will increase. Conversely, when the motor torque is less than that of the load, the speed drops.
当电机拖动机械负载时,转速通常是稳定的。在这种动平衡状态下,电动机产生的转矩 T 与负载转矩 T_L 大小相等方向相反。然而,当电动机转矩增加超过负载转矩时,转速增加。相反,当电动机转矩低于负载转矩时,转速下降。

2 With the load now running clockwise at a speed n_1, suppose we reduce T_M so that it is less than T_L. The net torque on the shaft now acts counterclockwise. Consequently, the speed decreases and will continue to decrease as long as T_L exceeds T_M.
当负载以转速 n_1 顺时针转动时,假设我们减小 T_M 使其小于 T_L。此时轴上的净转矩为逆时针方向。因此转速减小,只要 T_L 大于 T_M,转速将继续减小。

第三节　直流电机原理
Section 3　Principle of Direct-current Machines

白炽灯	incandescent lamp	调整曲线	regulation curve
并励	shunt-excited	叠片	lamination
波绕组	wave-winding	叠绕组	lap winding
槽	slot	定子	stator
差复励发电机	differential compound generator	端电压	terminal voltage
串励	series-excited	端盖	terminal box
磁极	field pole	短路	short-circuit
磁力线	lines of force	额定电压	rated voltage
等值电路	equivalent circuit	发电机	generator
电枢	armature	反电动势	counter-electromotive force
电枢反应	armature reaction	负载情况下	under load
电枢直径	armature diameter	负载特性	load characteristic
电压调整	voltage regulation	复励发电机	compound generator

滑环	slip ring	日光灯	fluorescent lamp
换向	commutation	剩磁	remnant flux
换向片	commutator segment	刷握	brush holder
换向片数	number of commutator bars	顺时针	clockwise
换向器	commutator	他励	separately excited
火花	sparking	炭刷	carbon brush
机械特性	torque-speed characteristic	铁芯	iron core
积复励发电机	over compound generator	温升	temperature rise
极数	number of poles	线圈	coil
绝缘	insulate	原理图	schematic diagram
开路	open-circuit	云母	mica
空载	no-load	匝数	number of turns
励磁电流	exciting current	整距	full-pitch
励磁绕组	field winding	整流器	rectifier
励磁线圈	field coil	直流电流	direct-current
满载	full-load	重载运行	heavy-duty operation
铭牌	nameplate	铸铁	cast iron
逆时针	counterclockwise	铸钢	cast steel
启动电流	starting current	转子	rotor
气隙	air gap	自感	self-inductance
绕组	winding	自励	self-excited

1 DC generators and motors are built in the same way; consequently, DC generator can operate as a motor and vice versa. Owing to their similar constructions, the fundamental properties of generators and motors are identical.

直流发电机和电动机以同样的方式制造；因此，直流发电机可以作为电动机运行，直流电动机也可以作为发电机运行。由于它们结构的相似性，发电机和电动机的基本特性是相同的。

2 Rheostat speed control: One way to control the speed of a DC motor is to place a rheostat in series with the armature. The current in the rheostat produces a voltage drop which subtracts from the fixed source voltage, which results in a smaller supply voltage across the armature. This method enables us to reduce the speed below its nominal speed.

可变电阻调速：一种控制直流电机转速的办法是电枢回路串可变电阻。流经可变电阻的电流产生压降，其对固定电源电压分压，结果电枢两端的电压降低。这种办法能够产生低于额定值的转速。

3 Among DC motors, the outstanding characteristics of each type are as follows. The series motor operates with a decidedly drooping speed as load is added, the no-load speed usually being prohibitively high; the torque is almost proportional to the square of the current at low flux levels. The shunt motor at constant field current operates at a slightly drooping but almost constant speed as load is added, the torque being almost proportional to armature current. Depending on the relative strengths of the shunt and series field, the compound motor is intermediate between the other two and may be given essentially the advantages of one or the other.

在直流电动机中，各种类型电机的主要特性描述如下：串励电动机运行时，随着负载的增加转速必然下降，通常空载转速会达到电机不允许的高度；转矩在磁通较小时几乎与电流的平方成正比。并励电动机在恒定励磁电流下运行时，随着负载的增加转速也会略微下降，但几乎是恒定的，转矩几乎与电枢电流成正比。复励电动机根据并励和串励绕组的相对强度，其特性介于两种电动机之间，本质上可以具有串励或并励电动机的优点。

第四节　他励直流电动机的运行
Section 4　Operation of Separately Excited DC Motors

泵　pump	恒转矩方式　constant torque mode
标称值　nominal value	恒转矩负载　constant torque load
补偿绕组　compensating winding	换向极　commutating pole
初始转速　initial speed	机电时间常数　mechanical time constant
磁通扭曲　flux distortion	机械功率　mechanical power
导体数　number of conductors	可变电阻　rheostat
电功率　electrical power	离心力　centrifugal force
电枢电流　armature current	励磁变阻器　field rheostat
电阻　resistor	摩擦损耗　friction loss
电阻值　resistance	能耗制动　dynamic braking
吊车　hoist	起动转矩　starting torque
额定负载　rated load	时间常数　time constant
额定功率　rated power	寿命　life expectancy
额定转速　rated speed	四象限运行　four-quadrant operation
反接制动　plugging	铁损耗　iron loss
反馈　feed back	铜损耗　copper loss
风扇　fan	温升　temperature rise
负载转矩　load torque	向心力　centripetal force
恒功率方式　constant horsepower mode	直线　straight line
恒功率负载　constant horsepower load	制动　braking

1 One is inclined to believe that stopping a DC motor is a simple, almost trivial, operation. Unfortunately, this is not always true. When a large DC motor is coupled to a heavy inertia load, it may take an hour or more for the system to come to a halt. For many reasons such a lengthy deceleration time is often unacceptable and, under these circumstances, we must apply a braking torque to ensure a rapid stop. One way to brake the motor is by simple mechanical friction, in the same way we stop a car. A more elegant method consists of circulating a reverse current in the armature, so as to brake the motor electrically.

人们总是觉得直流电动机的停机是一个简单甚至于微不足道的操作。遗憾的是并非总是如此。当一个大型直流电动机接惯性巨大的负载时,系统需要花费一个多小时的时间达到静止。由于多种原因这么长的减速时间是不可接受的,并且在这些情况下,我们必须加制动转矩以保证快速停机。制动电机的一个方法就是简单的机械摩擦,就像我们制动汽车时采用的方法一样。更适宜的方法是在电枢中加反向电流以实现电动机的电气制动。

2 Plugging: Plugging means that we suddenly reverse the armature current by reversing the terminals of the source and introduce a resistor R in series with the reversing circuit to limit the reverse current.

反接制动:反接制动指的是将电源端反接,突然使电枢电流反向,同时在电路中串入电阻 R 以限制反向电流。

3 We mentioned that the speed decreases exponentially with time when a DC motor is stopped by dynamic braking. We can, therefore, speak of a mechanical time constant T in much the same way we speak of the electrical time constant of a capacitor that discharges into a resistor. In essence, T is the time it takes for the speed of the motor to fall to 36.8 percent of its initial value.

我们提到过当直流电动机能耗制动停车时,转速随着时间的变化成指数级降低。因此谈到机械时间常数 T 时,与电容向电阻放电时的电时间常数基本相同。实际上,T 是电动机转速降到其初始值的 36.8% 时所用的时间。

第五节　变压器
Section 5　Transformer

安匝数　ampere-turns
按照　in terms of
变比　turns ratio
标幺值　per-unit value
并联　in parallel
初级线圈　primary winding
串联　in series
次级线圈　secondary winding
单相　single-phase
等值电路　equivalent circuit
低压边　low-voltage side
低压绕组　low-voltage winding
电度表　watt hour meter
电感　inductor
电感值　inductance
电抗　reactance
电流表　ammeter
电流互感器　current transformer
电流基值　base current
电容　capacitor
电容值　capacitance
电位计　potentiometer
电压表　voltmeter
电压互感器　voltage transformer
电压基值　base voltage
端子　terminal
发出功率　deliver power
反比于　be inversely proportional to
峰值　peak value
负相序　negative phase sequence
副边　secondary side
副边漏电抗　secondary leakage reactance
干式变压器　dry-type transformer
感性　inductive
高压边　high-voltage side
高压绕组　high-voltage winding
功率表　wattmeter
功率传输　power transmission
功率因数　power factor
共轭复数　complex conjugate
基值　base quantity
极性标注　polarity marking
降压变压器　step-down transformer

紧耦合　tight coupling
绝缘等级　insulation class
理想变压器　ideal transformer
领先　lead
漏磁通　leakage flux
落后　lag
耦合　coupling
配电变压器　distribution transformer
强迫通风冷却　forced-air cooled
容性　capacitive
三角形接法　delta connection
三铁心柱　3-legged core
三相　three-phase
三相对称电路　3-phase balanced circuit
升压变压器　step-up transformer
实部　real part
实际变压器　practical transformer
实值　actual value
视在功率　apparent power
瞬时功率　instantaneous power
松耦合　loose coupling
同名端　polarity-marked terminal
同相　in phase
无功功率　reactive power
无功功率表　varmeter
吸收功率　absorb power
线电流　line current
线电压　line voltage
相电流　phase current
相电压　phase voltage
相对于　with respect to
相角　phase angle
相量图　phasor diagram
相序　phase sequence
相移　phase shift
效率　efficiency
星型接法　wye connection
虚部　imaginary part
异相　out of phase
油浸　oil-immersed
油箱　oil tank
有功功率　active power
有效值　effective value

第三章 电机与拖动技术基础

原边　primary side
原边漏电抗　primary leakage reactance
杂散损耗　stray loss
折合　shift
正比于　be proportional to
正弦电流　sinusoidal current
正相序　positive sequence

中线　neutral line
主磁通　mutual flux
自冷　self-cooled
自耦变压器　autotransformer
阻抗　impedance
阻抗基值　base impedance

1 The transformer is probably one of the most useful electrical devices ever invented. It can raise or lower the voltage or current in an AC circuit. It can isolate circuits from each other, and it can increase or decrease the apparent value of a capacitor, an inductor, or a resistor.

变压器或许是已发明的最有用的电力设备之一。它可以提高或降低交流电路的电压或电流,隔离电路,增加或减少电容、电感或电阻的数值。

2 In the previous chapter we studied the ideal transformer and discovered its basic properties. However, in the real world transformers are not ideal and so our simple analysis must be modified to take this into account.

在前一章中我们学习了理想变压器并得出了其基本特性。然而,实际上变压器并非理想的。考虑到这一点,我们必须对简单的分析加以改进。

3 The short-circuit test can be used to find the equivalent series impedance $R_k + jX_k$. Although the choice of winding to short-circuit is arbitrary, for the sake of this discussion we will consider the short circuit to be applied to the transformer secondary and voltage applied to primary. For convenience, the high-voltage side is usually taken as the primary in this test. Because the equivalent series impedance in a typical transformer is relatively small, typically an applied primary voltage on the order of 10 to 15 percent or less of the rated value will result in rated current.

短路试验可用于求取等效串联阻抗 $R_k + jX_k$。虽然短路绕组的选择是任意的,但为了此处的讨论,将考虑在变压器二次侧短路而一次侧施加电压的情况。为方便起见,试验中通常取高压侧为一次侧。因为在典型变压器中,等效串联阻抗相对较小,一次侧电压一般加额定值的10%到15%或更小,就达到额定电流。

4 The open-circuit test is performed with the secondary open-circuited and rated voltage impressed on the primary. Under this condition an exciting current of a few percent of full-load current is obtained. Rated voltage is chosen to insure that the magnetizing reactance will be operating at a flux level close to that which will exist under normal operating conditions. For convenience, the low-voltage side is usually taken as the primary in this test.

开路试验通过将二次侧开路,在一次侧加额定电压来进行。在此条件下,得到的励磁电流为额定电流的百分之几。选择额定电压,是为了确保励磁电抗工作在与正常运行条件相近的某一磁通量级下。为方便起见,试验中通常取低压侧作为一次侧。

第六节　交流电机电枢绕组的电动势与磁通势
Section 6　EMF and MMF of Armature Winding of AC Machines

N极　N pole
S极　S pole
磁通势　magnetomotive force (MMF)
电动势　electromotive force (EMF)
电角度　electrical angle
叠绕组　lap winding
反转　reversing the direction of rotation
分解　decompose

感应　induction
感应电动机　induction motor
滑差　slip
机械角度　mechanical angle
基波　fundamental
基波频率　fundamental frequency
极距　pole pitch
节距　pitch

静止磁场　stationary field
均方根　root mean square (RMS)
每分钟转数　revolutions per minute
每极磁通　flux per pole
同步速　synchronous speed
线圈节距　coil pitch
相量　phasor
谐波　harmonic
谐波电压　harmonic voltage
行波　travelling wave

旋转磁场　revolving magnetic field/rotating field
旋转磁通　revolving flux
旋转方向　direction of rotation
异步电动机　asynchronous motor/induction motor
异步电机　asynchronous machine
右手定则　right-hand rule
正弦波　sine wave
转子导条　rotor bar
转子堵转　locked-rotor
左手定则　left-hand rule

1 The slip S of an induction motor is the difference between the synchronous speed and the rotor speed, expressed as a percent (or per-unit) of synchronous speed.
异步电动机的滑差 S 是同步转速与转子转速的差值，用同步转速的百分比来表示。

2 The rotational speed of the field depends upon the duration of one cycle, which in turn depends on the frequency of the source. If the frequency is 50 Hz, the resulting field makes one turn in 1/50 s, that is, 3000 revolutions per minute. Because the speed of the rotating field is necessarily synchronized with the frequency of the source, it is called synchronous speed.
磁场转速取决于一周的周期，后者又取决于电源频率。如果频率是 50 赫兹，所得磁场每 1/50 秒转一圈，即每分钟 3000 转。由于旋转磁场转速必须与电源频率同步，故称为同步转速。

3 In a typical machine, most of the flux produced by the stator and rotor windings crosses the air gap and links both windings; this is termed the mutual flux, directly analogous to the mutual or magnetizing flux in a transformer. However, some of the flux produced by the rotor and stator windings does not cross the air gap; this is analogous to the leakage flux in a transformer. These flux components are referred to as the rotor leakage flux and the stator leakage flux. Components of this leakage flux include slot and tooth tip leakage, end-turn leakage, and space harmonics in the air-gap field.
在典型的电机中，由定子绕组和转子绕组产生的磁通中的绝大部分都会穿越气隙，并同时链着两个绕组，这一磁通称为主磁通，这类似于变压器的主磁通或者励磁磁通。然而，定转子绕组产生的磁通中还有一部分不经过气隙，这类似于变压器中的漏磁通。这部分磁通称为转子漏磁通和定子漏磁通。这些漏磁通主要成分有槽和齿尖漏磁通、端部漏磁通和气隙磁场的空间谐波漏磁通等。

第七节　异步电动机原理
Section 7　Principles of Induction Motor

导条　bar
电磁功率　electromagnetic power
电磁转矩　electromagnetic torque
堵转实验　locked-rotor test
端环　end-ring
防爆　explosion-proof
防爆电动机　explosion-proof motor
防滴电动机　drip-proof motor
防溅电动机　splash-proof motor
风扇冷却　fan-cooled
风阻损耗　windage loss
过载　overload
滑动轴承　plain bearing

机壳　frame
开槽　punching
空载实验　no-load test
励磁电抗　magnetizing reactance
漏电抗　leakage reactance
内径　internal diameter
启动转矩　starting torque
球轴承　ball bearing
全封闭　totally-enclosed
绕线转子　wound rotor
绕线转子异步电动机　wound-rotor induction motor
鼠笼　squirrel-cage
鼠笼转子异步电动机　squirrel-cage induction motor

第三章 电机与拖动技术基础

凸极	salient-pole	隐极	cylindrical rotor
外径	external diameter	阻尼绕组	damper winding
无刷	brushless	最大转矩	maximum torque/breakdown torque

1 A squirrel-cage rotor is composed of bare copper bars, slightly longer than the rotor, which are pushed into the slots. The opposite ends are welded to two copper end-rings so that all the bars are short-circuited together. The entire construction (bars and end-rings) looks like a squirrel cage, from which the name is derived.

鼠笼转子是由比转子稍长一些的裸铜导条组成,它们被压入槽内。两端由两个铜端环焊接起来以形成短路。整个结构(导条和端环)类似一个鼠笼,"鼠笼"转子由此得名。

2 Although a wound-rotor motor costs more than a squirrel-cage motor, it offers the following advantages:

The locked-rotor current can be drastically reduced by inserting three external resistors in series with the rotor. Nevertheless, the locked-rotor torque will still be as high as that of a squirrel-cage motor. The speed can be varied by varying the external rotor resistors.

The motor is ideally suited to accelerate high-inertia loads, which require a long time to bring up to speed.

尽管绕线转子电动机比鼠笼转子电动机价格要高一些,它却有下列优点:
采用转子外串三个电阻可以极大地减小堵转电流。然而,堵转转矩却与鼠笼转子电动机一样高。
转速可以通过改变转子外串电阻来改变。
电动机非常适于加速惯性大的负载,后者需要长时间来加速。

3 The rotor terminals of an induction motor are short circuited by construction in the case of a squirrel-cage motor and externally in the case of a wound-rotor motor. The rotating air-gap flux induces slip-frequency voltages in the rotor windings. The rotor currents are then determined by the magnitudes of the induced voltages and the rotor impedance at slip frequency. At starting, the rotor is stationary, the slip is unity, and the rotor frequency equals the stator frequency. The field produced by the rotor currents therefore revolves at the same speed as the stator field, and a starting torque results, tending to turn the rotor in the direction of rotation of the stator-inducing field. If this torque is sufficient to overcome the opposition to rotation created by the shaft load, the motor will come up to its operation speed. The operating speed can never equal the synchronous speed however, since the rotor conductors would then be stationary with respect to the stator field. No current would be induced in them, and hence no torque would be produced.

感应电动机的转子端部被短路,在鼠笼型电动机中靠它的结构短路,在绕线式电动机中靠外部短路。旋转气隙磁通在转子绕组中感应转差频率的电势,转子电流由感应电势的大小和转差频率的转子阻抗决定。在启动时转子是静止的,转差率为1,转子频率等于定子频率,因此由转子电流产生的磁场以与定子磁场相同的转速旋转,产生启动转矩,并趋向于使转子沿着定子感应磁场的旋转方向旋转。如果这个转矩足够大,能克服轴上负载产生的阻力转矩,电动机将启动并达到它的运行速度。然而运行速度永远不可能等于同步速度,因为如果相等,转子导体就相对于定子磁场静止。在转子导体中不会产生感应电流,因此就不会产生转矩。

第八节 三相异步电动机的启动与制动
Section 8 Starting and Braking of 3-phase Induction Motors

变频器	frequency converter	单刀开关	single-pole switch
常闭触点	normally closed contact	第三象限	quadrant 3
常开触点	normally open contact	第一象限	quadrant 1
抽头	tap	电解电容	electrolytic capacitor
传导	conduction	断路器	circuit breaker

对流　convection
二极管　diode
辐射　radiation
隔离开关　disconnecting switch
过载继电器　overload relay
回馈制动　regenerative braking
继电器　relay
接触器　contactor
接地开关　grounding switch
可控硅　thyristor
框图　block diagram
逆变器　converter
热继电器　thermal relay

熔断器　fuse
三极管　transistor
三相四线制　3-phase 4-wire system
双刀双掷开关　double pole double throw switch
双速电机　two-speed motor
拖动转矩　driving torque
限位开关　limit switch
延时闭合　time-delay closing
延时断开　time-delay opening
制动转矩　braking torque
转矩转速曲线　torque-speed curve
自动开关　recloser
坐标　coordinate

1 The stator of a squirrel-cage induction motor can be designed so that the motor can operate at two different speeds. One way to obtain two speeds is to wind the stator with two separate windings having, say, 4 poles and 6 poles. The other way is to use special windings so that the speed is changed by simply changing the external stator connections.

鼠笼转子异步电动机的定子经过设计可使电动机运行于两个不同的转速。获得双速的一个办法是将定子绕成两个独立绕组，如4极和6极。另一个办法是使用特殊绕组，这样简单地改变定子外部的连接就可以改变转速。

2 The problem in controlling torque and speed is that the magnetizing current I_m and the torque-producing current I_2 are merged into a single current, namely the current I_1 flowing in the stator. In order to control the torque, this current must be split into its I_m and I_2 components. Furthermore, it is advantageous that I_m be held close to its rated value, to ensure the flux is as great as possible without excessive saturation of the iron.

控制转矩和转速的问题在于使励磁电流 I_m 和产生转矩的电流 I_2 合成为一个电流，即流入定子的电流 I_1。为了控制转矩，这个电流必须分成 I_m 和 I_2 分量。此外，保持 I_m 接近其额定值是有好处的，这使得铁芯未过分饱和时保证磁通尽可能大。

第九节　同步电动机
Section 9　Synchronous Motors

4极汽轮发电机　four-pole turbine generator
电动机惯例　motor reference direction
发电机惯例　generator reference direction
功角特性　power-angle characteristic
固态整流器　solid-state rectifier
过励磁　overexcitation
交轴　quadrature axis
矩角特性　torque-angle characteristic
空间基波分量　space-fundamental component
牵出同步　pull-out
牵出转矩　pull-out torque
牵入同步　pull-in
牵入转矩　pull-in torque
欠励磁　underexcitation

趋肤效应　skin effect
三次谐波磁通　third-harmonic flux
失步　loss of synchronism
水轮发电机　hydroelectric generator
同步电动机　synchronous motors
同步电感　synchronous inductance
同步电抗　synchronous reactance
凸极效应　effect of salient poles
无刷励磁系统　brushless excitation system
谐波效应　harmonic effect
隐极转子　cylindrical-rotor
原动机　prime mover
杂散损耗　stray loss
直轴　direct axis

1 A synchronous machine is one in which alternating current flows in the armature winding, and DC excitation is supplied to the field winding. The armature winding is almost invariably on the stator and is usually a three-phase winding. The field winding is on the rotor. The cylindrical-rotor construction is used for two- and four-pole turbine generators. The salient-pole construction is best adapted to multi-polar, slow-speed, hydroelectric generators and to most synchronous motors. The DC power required for excitation——approximately one to a few percent of the rating of the synchronous machine——is supplied by the excitation system.

同步电机是这样一种电机,其电枢绕组通过交流电流,而励磁绕组加直流电流。电枢绕组几乎无一例外地嵌放在定子上,而且通常为三相绕组。励磁绕组在转子上。隐级转子常用于2极或4极汽轮发电机;凸极结构转子特别适合于多极、低速水轮发电机和大多数同步电动机。励磁所需的直流功率大约为同步电机额定功率的百分之一到百分之几,这一功率由励磁系统提供。

2 The flux produced by an MMF wave in a uniform-air-gap machine is independent of the spatial alignment of the wave with respect to the field poles. In a salient-pole machine, however, the preferred direction of magnetization is determined by the protruding field poles. The permeance along the polar axis, commonly referred to as the rotor direct axis, is appreciably greater than that along the interpolar axis, commonly referred to as the rotor quadrature axis.

在均匀气隙电机中,磁势波所产生的磁通量,与其和磁极在空间是否对齐无关。但在凸极电机中,磁力线的首选走向取决于凸出的磁极。磁极的磁轴线通常称为转子直轴,两磁极间的中性线通常称为交轴。沿直轴的磁导显然大于沿交轴的磁导。

3 We have seen that a simple set of tests can be used to determine the significant parameters of a synchronous machine including the synchronous reactance. Two such tests are an open-circuit test, in which the machine terminal voltage is measured as a function of field current, and a short-circuit test, in which the armature is short-circuited and the short-circuit armature current is measured as a function of field current. These test methods are a variation of a testing technique applicable not only to synchronous machines but also to any electrical system whose behavior can be approximated by a linear equivalent circuit to which Thevenin's theorem applies.

我们已经知道,可以用一组简单的试验来确定同步电机的有效参数,包括同步电抗。其中的两个试验是开路试验和短路试验。开路试验可以测得电机端电压与励磁电流之间的函数关系;短路试验时电枢绕组短路,可以测得电枢短路电流与励磁电流之间的函数关系。这些试验方法是一种测试技术的特例。不仅对于同步电机,而且对于所有的能用戴维南线性等效电路近似表示其特性的电气设备都适用。

第十节 微控电机
Section 10 Micro-controlled Electrical Machines

步进电动机 stepper motor	马力 horsepower
电容启动 capacitor-start	脉振磁通势 pulsating MMF
电容运行 capacitor-run	伺服电动机 servo motor
电阻分相 resistance split-phase	特种电机 special motor
分马力 fractional horsepower	遥控 remote-control
分相 split-phase	原理图 schematic diagram
辅助绕组 auxiliary winding	罩极 shaded-pole
加速 accelerate	主绕组 main winding
减速 decelerate	自启动 self-start
交直流电动机 universal motor	阻尼 damping

1 Stepper motors are special motors that are used when motion and position have to be precisely controlled. As their name implies, stepper motors rotate in discrete steps, each step corresponding to a pulse that is supplied to one of its stator windings.

步进电机是特种电机,它们用于运动和位置需精确控制的场合。正如其名字所表明的,步进电机分步运转,每一步与加到定子绕组上的脉冲相对应。

2 Single-phase motors are the most familiar of all electric motors because they are used in home appliances and portable machine tools. In general, they are all employed when 3-phase power is not available. There are many kinds of single-phase motors on the market, each designed to meet a specific application.

由于单相电动机用于家用电器和便携式设备中,因此它们是所有电动机中最为人们所熟悉的。通常,在没有三相电源时使用单相电动机。市场上有多种单相电动机,每一种都设计用来满足特殊应用场合。

3 Capacitors can be used to improve motor starting performance, running performance, or both, depending on the size and connection of the capacitor. The capacitor-start motor is also a split-phase motor, but the time-phase displacement between the two currents is obtained by means of a capacitor in series with the auxiliary winding. Again the auxiliary winding is disconnected after the motor has started, and consequently the auxiliary winding and capacitor can be designed at minimum cost for intermittent service.

电容器可以被用来改进电动机的启动性能、运行性能或者同时改进两种性能。这取决于电容器的大小和连接。电容启动电动机也是分相电动机,只不过利用电容器和辅助绕组的串联来获得两个绕组电路之间的时间相位移。此外,当电动机启动后,辅助绕组被断开,因此,辅助绕组和电容器可以按间歇工作的最低成本来设计。

第四章 电力电子技术
Chapter 4　Power Electronics Technology

第一节　电力电子器件
Section 1　Power Electronics Device

MOS 控制晶闸管　MOS controlled thyristor
安全工作区　safe operating area (SOA)
半导体整流器　semiconductor rectifier
半控型器件　semi-control device
饱和区　saturated area
保护电路　protecting circuit
报警处理　alarm processing
报警电路　alarm circuit
背对背二极管　back-to-back diode
不可控器件　uncontrolled device
操作过电压　operating over voltage
场控晶闸管　field controlled thyristor
场效应管　field effect transistor
电力场效应晶体管　power MOSFET
电力电子器件　power electronic device
电力二极管　power diode
电力晶体管　giant transistor (GTR)
电容效应　capacitance effect
断态重复峰值电压　cut-off repeat peak voltage
额定电压　rate voltage
二次击穿　second breakdown
反向二极管　backward diode
反向二极晶闸管　reverse diode thyristor
反向恢复时间　reverse recovery time
反向击穿　reverse breakdown
反向截止状态　reverse cut-off state
反向偏置　reverse bias
反向偏置安全工作区　reverse biased safe operating area
反向重复峰值电压　reverse repeat peak voltage
反向阻断能力　reverse blocking capability
放大区　amplification area
负载电路　load circuit
高压集成电路　high voltage IC
功率集成电路　power integrated circuit (PIC)
汞弧整流器　mercury arc rectifier
固态继电器　solid state relay
关断电流　turn-off current
关断过程　cut-off process
关断过电压　cut-off over voltage
光控晶闸管　light triggered thyristor
过电流保护　over current protection
过电压保护　over voltage protection
换向过电压　commutation over voltage
恢复特性　recovery characteristic
集成电路　integrated circuit (IC)
集成门极换流晶闸管　integrated gate-commutated thyristor
检测电路　detecting circuit
截止区　cut-off area
晶闸管　thyristor
晶闸管的并联　thyristor parallel connection
晶闸管的串联　thyristor series connection
静电感应晶体管　static induction transistor (SIT)
静电感应晶闸管　static induction thyristor
静态特性　static characteristic
绝缘栅双极晶体管　insulated gate bipolar transistor (IGBT)
均流　mean current
均压　mean voltage
开启电压　turn-on voltage
开通过程　turn-on process
可控硅整流器　silicon controlled rectifier (SCR)
控制电路　control circuit
快恢复二极管　fast recovery diode
快恢复外延二极管　fast recovery epitaxial diode
快速晶闸管　fast switching thyristor
快速熔断器　fast acting fuse
浪涌电流　surge current
浪涌电压　surge voltage
零偏置　zero bias
门极　gate pole
门极关断开关　gate turn-off switch
门槛电压　gate voltage
逆导晶闸管　reverse conducting thyristor
逆向电流　backward current
漂移电流　floating current
平均通态电流　mean turn-on current

普通二极管 general purpose diode	万用表 universal meter
箝位 clamp	维持电流 holding current
箝位二极管 clamp diode	吸收电路 absorption circuit
强制风冷 air-blast cooling	吸收电容 absorbing capacitor
擎住电流 latching current	肖特基二极管 Schottky diode
擎住效应 latching effect	肖特基势垒二极管 Schottky barrier diode
驱动电路 drive circuit	续流元件 continuous current element
全控型器件 full control device	阴极 cathode
热击穿 hot breakdown	闸流管 thyratron
事故电流 abnormal current	正向电流 forward current
事故过电压 abnormal overload	正向偏置 forward bias
双极结型晶体管 bipolar junction transistor	正向偏置安全工作区 forward biased safe operating area
双向二极晶闸管 bi-directional diode thyristor	正向平均电流 forward mean current
双向晶闸管 bi-directional triode thyristor	正向压降 forward voltage drop
通态电流临界上升率 turn-on current critical climbing	智能功率模块 intelligent power module
通态电压 turn-on voltage	主电路 main circuit
通态电压临界上升率 turn-on voltage critical climbing	最高工作结温 maximum operation junction temperature

1 Thyristor: a generic term for gated power-switching semiconductors, including both SCRs and triacs.

晶闸管：其通用术语是门极控制开通的功率半导体器件，包括 SCR 和双向晶闸管。

2 Triac:

The triac is a device that contains two thyristors connected back to back. It has bi-directional voltage blocking capability and bi-directional current conduction capability. The triac is fully controlled from a single gate terminal. The triac has an approximately equal forward and reverse breakdown voltage and the gate triggers current conduction in both the first and the third quadrant with a low on-state voltage drop. A bi-directional thyristor is used to control the average current to an AC load; it differs from an SCR in that it can conduct current in either direction.

双向晶闸管：

双向晶闸管由两个背靠背晶闸管组成。它具有双向电压阻断能力和双向电流导通能力。双向晶闸管由一个门极控制端来控制，其正反向击穿电压和门极触发导通电流近似相等，在一、三象限通态压降较低。双向晶闸管用来控制交流负载的平均电流，它不同于 SCR 之处在于它可以流过双向电流。

3 Holding current is the amount of main-terminal current necessary to maintain an SCR or triac on conducting state once it has been fired.

保持电流是当 SCR 或双向晶闸管触发后，维持它们导通的主电流。

4 Protection:

Electronic switches in the power electronic converters are subjected to high rates of change in voltage and current that may rupture their physical structure and cause permanent damage. The switching devices must be protected from high transition conditions. Furthermore, the converters must be protected from the external surge conditions that may exist in the electric utility supply or may be acquired from other systems in the vicinity through radiation and electromagnetic interference.

保护：

电力电子转换电路中的电子开关承受很高的电压和电流变化率，这种高的变化率会使电子开关的物理结构发生变化从而引起器件的永久损坏。所以，必须对开关装置采取保护措施以防止频繁转换。更进一步说，转换电路必须设置保护，以防止存在由于电源的外部冲击或来自于附近其他系统的辐射和电磁干扰。

第四章 电力电子技术

第二节 整流电路
Section 2　Rectifier

半桥电路　half bridge circuit
贝克箝位电路　Baker clamping circuit
变流技术　power conversion technique
不控整流器　uncontrolled rectifier
超前角　angle of advance
触发　trigger
触发角　trigger angle
触发延迟角　trigger delay angle
单相半波调速系统　single-phase half-wave speed control system
单相半波可控整流电路　single-phase half-wave controlled rectifier
单相半桥逆变电路　single-phase half-bridge inverter
单相桥式全控整流电路　single-phase full-bridge controlled rectifier
单相全波可控整流电路　single-phase full-wave controlled rectifier
单相全桥逆变电路　single-phase full-bridge inverter
导通角　conduction angle
电气隔离　electrical isolation
环流　loop current
换向重叠角　commutation angle
基波因数　fundamental factor
畸变功率　distortion power
晶闸管相控变换器　thyristor phase-controlled converter
可控整流　controlled rectifier
漏感　leakage inductance
全波整流电路　full wave rectifier
全桥电路　full bridge circuit/full bridge converter
全桥整流电路　full bridge rectifier
三相半波可控整流电路　three-phase half-wave controlled rectifier
三相桥式可控整流电路　three-phase full-bridge controlled rectifier
同步整流电路　synchronous rectifier
相控　phase control
谐波　harmonic
移相全桥电路　phase shift full bridge converter
整流　rectification
整流二极管　rectifier diode
整流器　rectifier

1 Uncontrolled rectifier:
AC to DC converters without control are popularly known as rectifiers. The most commonly used AC source is 50 or 60 Hz voltage source which is available from the electric utility supply, also called the line source. The uncontrolled rectifiers are designed by using diodes. The designs are inexpensive and popular in industrial applications.

不控整流器:
最常见的不控交流—直流变换器是整流器。最常用的交流电源是电网供应的 50 或 60 赫兹的交流电压源,也称为线电压源。不控整流器采用二极管,这种设计成本低,工业应用很普遍。

2 AC to DC converter:
The AC to DC converter (rectifier) takes power from one or more AC voltage/current sources of single or multiple phases and delivers to a load. The output variable is a low-ripple DC voltage or DC current. Many practical AC-DC converters use a voltage source with line frequency of 50 or 60 Hertz, single-phase or three-phase. The AC-DC converters are widely used from very low-power battery chargers and DC power supply systems to very high-power DC motor drive systems.

交流—直流变换器:
交流—直流变换器(整流器)能量取之于一个或多个单相或多相的交流电压/电流源,并将其传送给负载。可控输出变量是低纹波直流电压或电流。许多实用的交—直变换器采用单相或三相,频率为 50~60 赫兹的电压源。交—直变换器广泛应用于小功率电池充电器、直流电源系统以及大功率直流电机传动系统。

3 Full-wave rectification:
A full-wave rectifier converts the whole of the input waveform to one of constant polarity (positive or

negative) at its output. Full-wave rectification converts both polarities of the input waveform to DC (direct current), and yields a higher mean output voltage. Two diodes and a center tapped transformer, or four diodes in a bridge configuration and any AC source (including a transformer without central tap), are needed. Single semiconductor diodes, double diodes with a common cathode or common anode, and four-diode bridges, are manufactured as single components.

全波整流：
全波整流器是将固定极性(正和负)的输入波形全部转换到其输出。全波整流将输入波形的两个极性转换成输出直流,并产生较高的平均输出电压。这种全波整流电路需要两个二极管和一个中心抽头变压器,或四个二极管组成桥及任何交流电源(包括无中心抽头的变压器)。单个半导体二极管,带有共阴或共阳的双二极管或者四个二极管的桥都可以被集成做成单个元件。

第三节　直流斩波电路
Section 3　DC Chopper

不可逆PWM变换器　irreversible PWM converter
冲击电压保护　surge protection
初始状态　initial condition
储能　accumulated energy
电流可逆斩波电路　current reversible chopper
电流连续模式　continuous conduction mode
断态(阻断状态)　off-state
反向击穿　reverse breakdown
辅助电源　accessory power supply
附属设备　accessory device
缓冲电路　snubber circuit
缓冲电容　snubber capacitor
击穿　breakdown
击穿电流　breakdown current
击穿特性　breakdown characteristic
降压斩波器　buck chopper/step down chopper
开关电源　switching mode power supply
开关关断特性　switching turn-off characteristic
开关瞬态过程　switching transient
开关损耗　switching loss
开关特性　switching characteristic
开关通态特性　switching turn-on characteristic

开关噪声　switching noise
开通　turn-on
可逆PWM变换器　reversible PWM converter
浪涌电压　surge voltage
脉冲幅度调制　pulse amplitude modulation
脉冲平均电路　pulse-averaging circuit
脉冲限幅　pulse clipping
脉冲载波　pulse carrier
脉动因数　pulsation
脉宽调制　pulse width modulation (PWM)
门极可关断晶闸管　gate turn-off thyristor
桥式可逆斩波电路　bridge reversible chopper
容许极限　acceptable limit
升降压变压器　buck and boost transformer
升压电路　boost up circuit
升压斩波电路　boost chopper/step up chopper
通态　on-state
通态损耗　turn-on loss
蓄电池　accumulator
直流斩波　DC chopping
直流斩波电路　DC chopping circuit
直流—直流变换器　DC-DC converter

1 Duty cycle is the percentage of the full cycle time spent in the up or on state.
占空比是开通时间占整个周期的百分数。

2 The DC-DC converters are widely used in the switching-mode power supplies and DC motor drive applications. Some of these converters, especially in power supplies, have an isolation transformer. The DC-DC converters are also used as interfaces between the DC systems of different voltage levels.
直流—直流变换器广泛应用于开关电源和直流电动机驱动。这些变换器特别是在电源中,要有隔离变压器。这种直流—直流变换器在不同电压等级的直流系统中被用作接口电路连接。

3 The DC-DC converter converts a DC power source to another DC source with different terminal specifications. The DC-DC converters change DC to AC first and then change AC back to DC. The DC source is often the DC voltage with ripple from AC to DC uncontrolled rectifier.

直流—直流变换器将直流电源转换成另一个具有不同参数的直流电源。直流—直流变换器先将直流变为交流,然后再将交流变回直流。直流电源通常采用交流变为直流的不控整流,其输出直流电压含有纹波。

4 The boost converter is a voltage step-up and current step-down converter. The current source is synthesized from a DC voltage source by adding a large value inductor in series. In DC regulated power supply applications, the voltage sink represents a large value capacitor in parallel with the load resistance. In the application of the DC motor control, the voltage sink represents the back EMF of the DC motor. The switching action produces a pulsating current. The capacitive filter smoothens the pulsating current and provides a DC voltage to the load.

升压斩波器是电压上升、电流下降的变换器。电流源是一个串联大电感的直流电压源,在直流稳压电源应用时,电压下降需要与负载电阻并联的大电容。在直流电机控制应用时,电压下降说明直流电机的电动势反向,开关作用可以产生脉冲电流,电容滤波可以平滑脉冲电流,给负载提供一个直流电压。

第四节 交流—交流变流电路
Section 4 AC - AC Converter Circuit

(电流)断续模式　discontinuous conduction mode　　交流调功电路　AC power controller
(电流)连续模式　continuous conduction mode　　　交流调压电路　AC voltage controller
电枢磁场　armature field　　　　　　　　　　　　晶闸管控制电抗器　thyristor controlled reactor
电枢控制　armature control　　　　　　　　　　　晶闸管投切电容器　thyristor switched capacitor
反并联　anti-parallel　　　　　　　　　　　　　　矩阵式变频器　matrix converter
交交变频电路　AC-AC frequency converter　　　　周波变换器/交交变频　cycloconverter
交流电力电子开关　electronic AC switch

1 AC-AC converters transfer power from one AC system to another with waveforms of different amplitude, frequency, or phase. These converters are designed for one-quadrant to four-quadrant operation. Typical applications are variable speed drives and four-quadrant PWM drive for traction.

交流—交流变换器是将一个交流系统变成另一个不同幅值、频率或相位的变换电路。这种变换器可以把单相变为四相电源,典型应用是变频调速和 PWM 四象限拖动系统。

2 Cycloconverter drives:
The cycloconverter output is derived from the AC line input without going through a DC link. The output frequency is lower than the input frequency (usually less than one-third). Cycloconverter drivers are used in a low speed and very large horsepower.

交交变频传动:
交交变频传动输出来自于交流线性输入,没有直流连接环节。输出频率比输入频率低(通常低于三分之一)。交交变频用于低速大功率场合。

3 The AC controller controls the amplitude of the AC source waveform. The amplitude can be controlled by several methods:
①Turning the converter on and off periodically and controlling the on or off duty cycle;
②Controlling the delay angle, that is, the angle of turning on of switches, or
③Switching at a high frequency by a pulse width modulated waveform.

交流控制器控制调节交流电源电压的幅值,可以通过以下方法控制交流电压的幅值:
①周期打开和关断转换装置,控制其开通或关断与周期的比值;
②控制延迟角,也就是开关器件的导通角;或者
③通过脉宽调制的方法控制高频开关。

第五节　逆变电路
Section 5　Inversion Circuit

安全系数　assurance factor
低电压纹波　low voltage ripple
电流(源)型逆变电路　current source inverter
电网换流　line commutation
电压(源)型逆变电路　voltage source inverter
多电平逆变电路　multi-level inverter
多重化　multiplex
多重逆变电路　multiplex inverter
反激电路　flyback converter
反向控制　reversed control
反转　back run
峰—峰值　peak-to-peak value
负载换流　load commutation
换流　commutation

交流鼠笼电机　AC squirrel-cage motor
矩阵式变频电路　matrix frequency converter
可逆循环　reversible cycle
逆变　inversion
器件换流　device commutation
强迫换流　forced commutation
无源逆变　reactive invertion
有源功率滤波器　active power filter
有源逆变　regenerative invertion
有源器件　active device
阈值电压　threshold voltage
中性点箝位型逆变电路　neutral point clamped inverter

1 Commutation refers to the interruption of main terminal current within the SCR by connecting a temporary short circuit from the anode to the cathode, used for turning off an SCR in a DC circuit.
换流是指在直流电路中,为了使晶闸管关断,当主电流断续时,在 SCR 阳极和阴极之间形成暂时的短路使主电流中断。

2 Inverter drive refers to a variable frequency motor-drive circuit that converts DC into AC motor voltage, as opposed to performing an AC-to-AC conversion.
逆变驱动:将直流电压变为给变频传动电机送电的交流电压电路,在性能上与交流变交流转换电路完全不同。

3 DC-AC converters:
DC-AC converters are conventionally called the inverters. Such converters are very popular in the battery operated power systems such as the un-interruptible power supplies(UPS)for hospitals, and AC motor drive. Low power level inverter are usually single phase type and medium and high power inverters are three phase type. We will look into the performance of DC-AC converters of several useful topologies. The focus will be on the inverter operation for low frequency output voltages such as 50-60Hz or 400Hz in aircraft systems. Analyses of DC-AC converter circuits at low and high frequencies are very similar except that at high frequencies the parasitic capacitances and inductances, charge storage, and heat localizing problems in switching devices must be taken into account.
直流—交流变换器:
直流—交流变换器通常被称为逆变器。这种变换器常用在为医院和交流电机驱动等供电的不间断电源中,作为配套电源。低电压逆变器采用单相逆变器,而中高电压逆变器采用三相逆变器,我们通过几种常用变换器的结构可以看出其特性,变换器工作的关键是输出 50~60 赫兹的低频电压及航空系统的 400 赫兹输出电压。除在高频时必须考虑附加电容和电感、电荷储存及开关内的热聚集问题外,低频和高频情况下直流—交流变换器的分析非常相似。

第六节　PWM 控制技术
Section 6　PWM Control Technology

采样周期　sampling period
调制波　modulating wave
调制度　modulating ratio
方波　square wave
功率模块　power module
功率因数　power factor
功率因数校正　power factor correction
关断　turn-off
规则采样法　rule sampling method
锯齿波调制　serrasoid modulation
控制模式　control pattern
脉冲后沿　pulse tail
脉冲宽度　pulse width

脉冲宽度 幅度变换器 pulse width-amplitude converter
脉冲前沿　pulse front
三角波　triangular wave
特定谐波消去法　selected harmonic elimination
通态（导通状态）　on-state
同步调制　synchronous modulation
异步调制　asynchronous modulation
载波　carrier wave
载波比　carrier wave ratio
载波频率　carrier wave frequency
正弦 PWM　sinusoidal PWM(SPWM)
自然采样法　natural sampling method

1 Modulation is the technique of continuously varying the average power to an electrical load by varying the duty cycle (width) of pulses applied to the load.

调制是指通过改变负载上脉冲的占空比来连续改变平均功率的技术。

2 PWM control：

Phase control of AC-AC converters generates low order harmonics. The unwanted low order harmonics are difficult to reduce. The problem can be overcome by multiplying the AC voltage source waveform with a high frequency pulse train. The amplitude of the output voltage is controlled by varying the pulse width. The switching frequency must be an integral multiple of the source frequency.

脉宽调制控制：

交流—交流变换器的相位控制产生低次谐波。这些有害的低次谐波很难消除。解决这个难题的方法是采用多重与交流电压源波形相同的高频脉冲。输出电压的幅值由调整脉冲宽度控制,开关频率必须是电源频率的整数倍。

3 PWM is used in efficient voltage regulators. By switching voltage to the load with the appropriate duty cycle, the output will approximate a voltage at the desired level. The switching noise is usually filtered with an inductor and a capacitor. One method measures the output voltage. When it is lower than the desired voltage, it turns on the switch. When the output voltage is above the desired voltage, it turns off the switch.

PWM 技术用来调节电压很有效,通过适当的占空比来控制开关电压,负载输出将在所期望的电压值附近。开关噪声通常通过电感和电容滤掉。测量输出电压值,当低于期望值时,打开开关,当高于期望值时,关掉开关。

第七节　软开关技术
Section 7　Soft Switching Technology

并联谐振式逆变电路　parallel-resonant inverter
断态　off-state
静止无功补偿器　static var compensator(SVC)
零电流　zero current
零电流开关准谐振电路　zero-current-switching quasi-resonant converter

零电流转换 PWM 电路　zero current transition PWM converter
零电压　zero voltage
零电压开关多谐振电路　zero-voltage-switching multi-resonant converter
零电压开关准谐振电路　zero-voltage-switching

quasi-resonant converter
零电压转换 PWM 电路　zero voltage transition PWM converter
零开关　zero switching
零转换　zero transition
脉冲频率调制　pulse frequency modulation

软开关　soft switching
通态　on-state
谐振　resonation
谐振直流环电路　resonant DC link
硬开关　hard switching
准谐振　quasi-resonant

1 Zero-current switching:

Turning on and off a switch at zero current is the surest way of minimizing switching losses. An inductor in series with the switch will ensure zero current turn-on because the current through the inductor cannot build instantaneously. However, the series inductor makes turnoff at zero current impossible.

零电流开关:

开关在零电流时开、关是减少开关损耗最佳方式。串联在开关电路中的电感可以保证零电流开通，因为通过电感的电流不能瞬间建立，但串联电感在零电流不可能关断。

2 Zero-voltage switching:

Turning on and off a switch at zero voltage is the surest way of minimizing switching losses. A capacitor in parallel with the switch will ensure zero voltage turn-on because the voltage through the capacitor cannot build instantaneously. However, the parallel capacitor makes turnoff at zero voltage impossible.

零电压开关:

开关在零电压时开、关是减少开关损耗的最佳方式。并联在开关电路中的电容可以保证零电压开通，因为电容的电压不能瞬间建立，但并联电容在零电压不可能关断。

3 Switching losses:

Power losses in the power electronic converters are comprised of the switching losses and the parasitic losses. The parasitic losses account for the losses due to the winding resistances of the inductors and transformers, the dielectric losses of capacitors, the eddy and the hysteresis losses. The switching losses are significant and can be further divided into three components: ①the on-state losses, ②the off-state losses and ③the losses in the transition states.

开关损耗:

在电力电子变换器的功率损失包含开关损耗和寄生损耗。寄生损耗是与电感器和变压器的绕线电阻、电容的介质损耗、涡流和磁滞损失这些量有关的。开关损耗的意义很明显，可以划分成三个部分: ①通态损耗,②断态损耗,③过渡状态的损耗。

第八节　组合变流电路
Section 8　Combination Converter

变压变频　variable voltage variable frequency (VVVF)
不间断电源　uninterruptible power supply (UPS)
单端电路　single end circuit
电力变换　power conversion
电力电子技术　power electronic technology
电力电子系统　power electronic system
电力电子学　power electronics
电力系统分析　power system analysis
反激电路　flyback circuit
供电电压　power-supply voltage
恒压恒频　constant voltage constant frequency (CVCF)

间接电流控制　indirect current control
间接直流变换电路　indirect DC-DC converter
开关电源　switching mode power supply
双端电路　double end circuit
推挽变流器　push-pull converter
正激电路　forward circuit
直交直电路　DC-AC-DC converter
直接电流控制　direct current control
直接直流变换电路　direct DC-DC converter
自动闭锁信号　automatic block signal
自动控制牵引设备　automatic traction equipment

第四章 电力电子技术

1 The AC source is first converted into a DC voltage or a DC current source. The DC source is then converted into the AC source of the desired specifications in amplitude, frequency and phase. The intermediate DC stage is called the DC link or the dc bus. The output voltage in the DC-link converter may have a different amplitude, a different frequency and a different number of phases from the input voltage source.

交流电源首先被转换成直流电压或直流电流源,然后再转换成幅值、频率和相数符合要求的交流电源。这种中间的直流环节被称为直流连接或直流总线。直流连接的逆变器的输出电压与输入电压相比,其幅值、频率以及相数都不同。

2 Full-bridge converter:
The full-bridge topology constitutes the four-quadrant DC-DC converter. The converter permits the output voltage of both polarities and each one with positive and negative polarity currents. Power flows in both directions. A typical application is the forward and reverse motion with dynamic baking of DC motors, such as in the elevator drive.

全桥变换器:
全桥电路由四象限直流-直流变换器构成。变换器的电流和电压都是双极性的。功率可以双向流动。典型应用是可以动态制动直流电机的正反向运行,例如电梯的运行。

3 Emergency facilities require uninterrupted AC power supplies as protection against power outages and over-voltage and under-voltage situations from line transients and harmonic disturbances. They are, in general, not long-term sources. They are available to provide power until the backbone supply is restored.

由于电网瞬时和谐波干扰产生的电源断电和过电压、低电压情况下的保护措施,需要不间断交流电源作为应急设施。一般情况下,这种电源不是长时工作,在主电源恢复正常工作之前,由它提供电源供电。

第五章 电路原理
Chapter 5　Principle of Electric Circuits

第一节　电路模型和电路定律
Section 1　Circuit Model and Circuit Law

u-i 特性　u-i characteristic
安培　ampere (A)
参考方向　reference direction
超导体　superconductor
代数和　algebraic sum
电导　conductance
电荷　charge
电流　current
电流表　ammeter
电路　electric circuit
电路参数　circuit parameter
电路符号　circuit symbol
电路模型　circuit model
电路图　circuit diagram
电路元件　circuit element
电位　potential
电位参考点　potential reference point
电位差(电势差)　potential difference
电位降　potential drop
电位升　potential rise
电压　voltage
电压表　voltmeter
电压电流关系　voltage current relation
电源　source
电子　electron
电阻元件　resistor
短路　short circuit
非线性电路　nonlinear circuit
非线性电阻　nonlinear resistance
非线性电阻元件　nonlinear resistor
伏特　volt(V)

负电荷　negative charge
功率　power
关联参考方向(无源符号约定)　passive sign convention
国际单位制　international system of units (SI)
赫兹(赫)　hertz(Hz)
基尔霍夫电流定律　Kirchhoff's current law(KCL)
基尔霍夫电压定律　Kirchhoff's voltage law(KVL)
基尔霍夫定律　Kirchhoff's laws
焦耳　Joule(J)
开路　open circuit
库仑　coulomb
零电位点　zero potential point
能量　energy
欧姆　Ohm
欧姆定律　Ohm's law
千赫　kilohertz
输出　output
输入　input
瓦特　watt(W)
无源元件　passive element
西门子　Siemens(S)
线性电路　linear circuit
线性电阻　linear resistance
线性电阻元件　linear resistor
有源元件　active element
元件　element
正电荷　positive charge
直流电流源　DC current source
直流电压源　DC voltage source

1 An electric circuit consists of electrical elements connected together.

电子电路由若干个相互连接在一起的电路元件组成。

2 Current is the rate of charge flow.

$$i=\frac{dq}{dt}$$

电流是电荷流动的变化率。

$$i=\frac{dq}{dt}$$

第五章　电路原理

3 Voltage is the energy required to move 1C of charge through an element.
$$v=\frac{dw}{dq}$$
电压是移动 1C 电荷流过元件所需要的能量。
$$v=\frac{dw}{dq}$$

4 Power is the energy supplied or absorbed per unit time. It is also the product of voltage and current.
$$p=\frac{dw}{dt}=vi$$
功率是单位时间所提供(发出)或吸收的能量,也可表达为电压与电流的乘积。
$$p=\frac{dw}{dt}=vi$$

5 According to the passive sign convention, power assumes a positive sign when the current enters the positive polarity of the voltage across an element.
无源符号规则规定,若电流由接电压正端处流入元件,则功率的符号为正。

6 Ohm's law states that the voltage across a resistor is directly proportional to the current flowing through the resistor.
欧姆定律:电阻两端的电压与流经该电阻的电流成正比。

7 An ideal independent source is an active element that provides a specified voltage or current that is completely independent of other circuit variables. An ideal dependent (or controlled) source is an active element in which the source quantity is controlled by another voltage or current.
理想独立电源是有源(电源)元件,它独立地提供一定的电压或电流,与电路中的其他参量无关。理想非独立源(受控源)是有源元件,它所提供的电压或电流是受别的电压或电流控制的。

8 Kirchhoff's current law states that the algebraic sum of all the currents at any node in a circuit equals zero. In other words, the sum of the currents entering a node is equal to the sum of the currents leaving the node.
基尔霍夫电流定律:在电路中,任何结点上的所有电流的代数和等于零。或者说,流入结点电流之和等于流出该结点电流之和。

9 Kirchhoff's voltage law states that the algebraic sum of all the voltages around any closed path in a circuit equals zero. In other words, the sum of voltage rises equals the sum of voltage drops.
基尔霍夫电压定律:在电路中,环绕任何闭合路径的所有电压的代数和等于零。或者说,电压升高之和等于电压降落之和。

10 Section Objectives:
(1) Learn about the basic assumptions of circuits analysis;
(2) Understand the behavior of each ideal circuit element in terms of its voltage and current;
(3) Be familiar with the passive sign convention;
(4) Be familiar with Kirchhoff's current and voltage laws.
本节学习目标:
(1)了解电路分析的基本假设;
(2)理解各理想电路元件的伏安特性;
(3)熟练掌握关联参考方向;
(4)熟练掌握基尔霍夫电流和电压定律。

第二节　电阻电路的等效变换
Section 2　Equivalent Transformation of Resistive Circuits

Δ—Y 变换　delta-to-wye conversion
并联电路分析　analysis of parallel circuit
并联连接的电路元件　parallel-connected circuit elements

串并联电路　series-parallel circuit
串并联电路分析　analysis of series-parallel circuits
串联电路分析　analysis of series circuit
串联连接的电路元件　series-connected circuit elements
等效串联电阻　equivalent series resistance
等效电导　equivalent conductance
等效电阻　equivalent resistance
电路分析　circuit analysis
电路理论　circuit theory
电压表的负载效应　voltmeter loading effects
电源变换　source transformation
电阻的并联　resistors in parallel
电阻的串联　resistors in series
电阻电路　resistive circuit
二端网络　two-terminal network
二端元件　two-terminal element
分流电路　current-divider circuit
分压电路　voltage-divider circuit
集中参数　lumped parameter
集中参数电路　lumped circuit
集中参数元件　lumped element
交流电源　alternating current (AC) source
理想电路元件　ideal circuit element
三端元件　three-terminal element
三角形电阻网络　delta-connected resistance network
输入电阻　input resistance
星形电阻网络　star-connected resistance network
直流电源　direct current (DC) source

1 A linear network consists of linear elements, linear dependent sources, and linear independent sources, etc.
一个线性网络由线性元件、线性受控源和线性独立源等组成。

2 Two elements are in series when they are connected sequentially, end to end. When elements are in series, the same current flows through them. They are in parallel if they are connected to the same two nodes. Elements in parallel always have the same voltage across them.
头尾依次连接的两个元件称为串联，流过串联元件的电流相同。两个元件的两端连到相同的两个节点上称为并联，并联元件两端的电压相同。

3 Resistors in series behave as a single resistor whose resistance equals to the sum of the resistances of the individual resistors. Conductances in parallel behave as a single conductance whose value is equal to the sum of the individual conductances.
多个电阻串联的等效电阻值等于各个电阻值的和。多个电阻并联的等效电导等于各个电导之和。

4 A source transformation is the process of replacing a voltage source in series with a resistor by a current source in parallel with a resistor, or vice versa.
电源变换是用带并联电阻的电流源取代带串联电阻的电压源(或者反之)的一种转换过程。

5 Section Objectives:
(1) Learn about combining resistors in series and in parallel.
(2) Learn about wye-delta transformation.
(3) Become familiar with the source transformation.
(4) Learn about methods used for simplifying analysis.
本节学习目标：
(1)了解电阻的串并联；
(2)了解星—三角变换；
(3)掌握电源变换；
(4)学习简化电路的方法。

第三节　电阻电路的一般分析
Section 3　Methods of General Analysis in Resistive Circuits

并联电路　parallel circuit
参考结点　reference node
串联电路　series circuit
对称网络　bilateral network
非平面电路　nonplanar circuits
附加方程　supplemented equation

第五章　电路原理

广义结点	supernode	平面电路	planar circuits
广义网孔	supermesh	树	tree
互导	mutual conductance	树支	tree branch
互阻	mutual resistance	图	graph
回路	loop	网孔	mesh
回路电流	loop current	网孔电流	mesh current
回路电流法	loop-current method	网孔电流法	mesh-current method
回路分析法	loop analysis method	网孔分析法	mesh analysis
节点	node	有伴电流源	accompanied current source
结点电压	node voltage	有伴电压源	accompanied voltage source
结点电压法	node-voltage method	支路	branch
结点分析法	nodal analysis method	支路电流	branch current
连通图	connected graph	支路电流法	branch-current method
连支	link	支路电压	branch voltage
路径	path		

1 A branch is a single two-terminal element in an electric circuit. A node is the point of connection between two or more branches. A loop is a closed path in a circuit. The number of branches b, the number of nodes n, and the number of independent loops l in a network are related as
$$b=l+n-1$$
支路是电路中的单个二端元件。结点是两个及两个以上支路的连接点。电路中的闭合路径叫回路。网络中的支路数 b、结点数 n 和独立回路数 l 满足如下关系：
$$b=l+n-1$$

2 Mesh-current method is the application of Kirchhoff's voltage law around meshes in a planar circuit. We express the result in terms of mesh currents. Solving the simultaneous equations yields the mesh currents.
网孔电流法是平面电路中应用基尔霍夫电压定律绕网孔一周的方法，其结果用网孔电流表示。通过解联立方程组得到各个网孔电流。

3 Node-voltage method is the application of Kirchhoff's current law at the nonreference nodes (It is applicable to both planar and nonplanar circuits.). We express the result in terms of the node voltages. Solving the simultaneous equations yields the node voltages.
结点电压法是基尔霍夫电流定律在非参考结点上的应用（它可应用于平面电路和非平面电路），其结果用结点电压表示。通过解联立方程组得到各个结点的电压。

4 Node-voltage method is normally used when a circuit has fewer node equations than mesh equations. Mesh-current method is normally used when a circuit has fewer mesh equations than node equations.
结点电压法一般用于电路中结点方程比网孔方程少的情况。网孔电流法一般应用于电路中网孔方程比结点方程少的情况。

5 Section Objectives：
(1) Understand the concepts of nodes, branches and loops.
(2) Understand the concepts of supernode and supermesh.
(3) Learn about node-voltage and mesh-current methods.
本节学习目标：
(1) 理解结点、支路和回路的概念；
(2) 理解广义结点和广义网孔的概念；
(3) 学习结点电压法和网孔电流法。

第四节 电路定理
Section 4　Circuit Theorems

原理　principle	dependent source/controlled source
定理　theorem	负载　load
定律　law	负载效应　loading effect
戴维宁等效电路　Thévenin equivalent circuit	工作电压　working voltage
戴维宁定理　Thévenin's theorem	互易定理　reciprocity theorem
等效电路　equivalent circuit	互易条件　reciprocity condition
等效发电机　equivalent generator	互易网络　reciprocity network
电流控制电流源（CCCS）　current controlled current source (CCCS)	激励　excitation
	诺顿等效电路　Norton equivalent circuit
电流控制电压源（CCVS）　current controlled voltage source (CCVS)	诺顿定理　Norton's theorem
	匹配　matching
电流源　current source	特勒根定理　Tellegen's theorem
电压控制电流源（VCCS）　voltage controlled current source (VCCS)	特勒根功率定理　Tellegen's power theorem
	特勒根似功率定理　Tellegen's quasi-power theorem
电压控制电压源（VCVS）　voltage controlled voltage source (VCVS)	替代定理　substitution theorem
	线性电路　linear circuit
电压源　voltage source	线性对称电路　linear bilateral circuit
电压源内阻　internal resistance of voltage source	线性值　linear quantity
叠加原理　superposition principle	线性组合　linear combination
独立电源　independent source	原电路　original circuit
对外等效　equivalent external to the conversion	最大功率传输　maximum power transfer
二端电路　two-terminal circuit	
非独立(受控)电源	

1 Network theorems are used to reduce a complex circuit to a simpler one, thereby making circuit analysis much simpler.

网络的若干定理用于将复杂电路转化为简单电路,从而使电路分析更为简易。

2 The superposition principle states that the voltage across (or current through) an element in a linear circuit is the algebraic sum of the voltages across (or currents through) that element due to each independent source acting alone.

叠加原理:线性电路中,任一器件上的电压或电流都是电路中各个独立电源单独作用时在该处产生的电压或电流的叠加。

3 Substitution theorem states that any branch within a circuit may be replaced by an equivalent branch, provided the replacement branch has the same current through it and voltage across it as the original branch.

替代定理:给定电路的任一支路都可由等效支路来替代,替代支路的电流和电压均与原支路的值相同。

4 Thévenin's theorem states that a linear two-terminal circuit can be replaced by an equivalent circuit consisting of a voltage source V_{Th} in series with a resistor R_{Th}, where V_{Th} is the open-circuit voltage at the terminals and R_{Th} is the input or equivalent resistance at the terminals when the independent sources are turned off.

戴维宁定理:线性二端电路可以由一个等效电路来取代,该电路由一个电压源 V_{Th} 与一个电阻 R_{Th} 串联组成。V_{Th} 等于二端电路的开路电压,R_{Th} 等于独立电源置零后的输入(或等效)电阻。

5 Norton's theorem states that a linear two-terminal circuit can be replaced by an equivalent circuit consisting of a current source I_N in parallel with a resistor R_N, where I_N is the short-circuit current through the terminals and R_N is the input or equivalent resistance at the terminals when the independent sources are turned off.

诺顿定理：线性二端电路可以被由一个电流源 I_N 与一个电阻 R_N 并联组成的等效电路来取代，I_N 等于流过端点的短路电流，R_N 等于独立电源置零后的输入（或等效）电阻。

6 For a given Thévenin equivalent circuit, maximum power transfer occurs when $R_L = R_{Th}$, that is, when the load resistance is equal to the Thévenin resistance.

对于一个给定的戴维宁等效电路，当负载电阻等于戴维宁电阻时，即 $R_L = R_{Th}$，传递到负载上的功率最大。

7 Section Objectives：
(1) Understand the principle of superposition.
(2) Become familiar with Thevenin's and Norton's theorems.
(3) Understand the concept of maximum power transfer to the load.

本节学习目标：
(1) 理解叠加原理；
(2) 熟练掌握戴维宁和诺顿定理；
(3) 理解负载最大功率传输的概念。

第五节 含有运算放大器的电阻电路
Section 5 Resistive Circuits with Operational Amplifiers

差动输入　differential mode input
差动输入电压　differential input voltage
倒向比例电路　inverting-amplifier circuit
倒向放大器　inverting amplifier
倒向输入端　inverting input terminal
电压传输特性　voltage transfer characteristic
电压跟随器　voltage follower
反相输入　inverting input
放大电路　amplifier circuit
负电源　negative power supply
负反馈　negative feedback
公共端　common node
集成电路　integrated circuit
加法电路　summing-amplifier circuit
晶体管　transistor
开环电压增益　open-loop voltage gain
开环放大倍数　open-loop gain
理想运算放大器　ideal op amp
输出电压　output voltage
输入电压　input voltage
输入阻抗　input impedance
同相输入　noninverting input
同相输入端　noninverting input terminal
线性区　linear region
虚短　virtual short
运算放大器　operational amplifier(op amp)
增益　gain
正电源　positive power supply

1 An op amp is an active circuit element designed to perform mathematical operations of addition, subtraction, multiplication, division, differentiation, and integration.
运算放大器是一个有源电路器件，用于进行诸如加减乘除、微分、积分等运算。

2 The op amp is a high-gain amplifier that has high input resistance and low output resistance.
运算放大器是一个高增益放大器，其输入电阻很大，而输出电阻很小。

3 An ideal op amp is an amplifier with infinite open-loop gain, infinite input resistance, and zero output resistance.
理想运算放大器是开环增益为无穷大、输入电阻为无穷大、输出电阻为零的运算放大器。

4 Two important characteristics of the ideal op amp are：
(1) The currents into both input terminals are zero.
(2) The voltage across the input terminals is negligibly small.
理想运算放大器有两个重要特性：

(1)流入两个输入端的电流均为零；
(2)两个输入端之间的电压很小，可以忽略不计。

5 In an inverting amplifier, the output voltage is a negative multiple of the input.
反相放大器中输出电压是输入电压乘以负的倍数。

6 In a noninverting amplifier, the output is a positive multiple of the input.
同相放大器中输出电压是输入电压乘以正的倍数。

7 In a voltage follower, the output follows the input.
电压跟随器：其输出电压等于(跟随)输入电压。

8 Op amp circuits may be cascaded without changing their input-output relationships.
运算放大器电路可以级联，且不影响自身的输入—输出关系。

9 Section Objectives：
(1) Become familiar with the circuit model of op amp.
(2) Learn how to analyze circuits involving an ideal op amp.
本节学习目标：
(1)掌握运算放大器的电路模型；
(2)学会含有理想运算放大器的电路的分析。

第六节 储能元件
Section 6　Energy-storage Elements

储能元件　storage element/energy-storage element	亨利　henry
电感　inductance	记忆元件　memory element
电感元件　inductor	楞次定律　Lenz's law
电容　capacitance	皮法　picofarad
电容元件　capacitor	微法　microfarad
电通量　electric flux	微亨　microhenry
动态元件　dynamic elements	无记忆元件　memoryless element
法拉　farad	无源电路元件　passive circuit elements
法拉第电磁感应定律	无源线性元件　passive linear elements
Farady's law of electromagnetic	线性电感元件　linear inductor
非线性电感元件　nonlinear inductor	线性电容元件　linear capacitor
非线性电容元件　nonlinear capacitor	线性元件　linear elements
毫亨　millihenry	右手螺旋定则　right-handed screw rule

1 The current through a capacitor is directly proportional to the time rate of change of the voltage across it.

$$i = C\frac{dv}{dt}$$

The current through a capacitor is zero unless the voltage is changing. Thus, a capacitor acts like an open circuit to a DC source.
流经电容器的电流直接正比于其两端电压随时间的变化率：

$$i = C\frac{dv}{dt}$$

除非电压随时间正在变化，否则流过电容器的电流为零。所以对直流电源而言，电容器的作用相当于开路。

2 The voltage across a capacitor is directly proportional to the time integral of the current through it.

$$v = \frac{1}{C}\int_{-\infty}^{t} i\,dt = \frac{1}{C}\int_{t_0}^{t} i\,dt + i(t_0)$$

The voltage across a capacitor cannot change instantly.

电容两端的电压直接正比于流过它的电流对时间的积分：

$$v = \frac{1}{C}\int_{-\infty}^{t} i\, dt = \frac{1}{C}\int_{t_0}^{t} i\, dt + i(t_0)$$

电容器两端的电压不能突变。

3 The voltage across an inductor is directly proportional to the time rate of change of the current through it.

$$v = L\frac{di}{dt}$$

The voltage across the inductor is zero unless the current is changing. Thus an inductor acts like a short circuit to a DC source.

电感器两端的电压直接正比于流过它的电流随时间的变化率：

$$v = L\frac{di}{dt}$$

除非电流随时间正在变化，否则电感两端的电压为零。所以对直流电源而言，电感器的作用相当于短路。

4 The current through an inductor is directly proportional to the time integral of the voltage across it.

$$i = \frac{1}{L}\int_{-\infty}^{t} v\, dt = \frac{1}{L}\int_{t_0}^{t} v\, dt + v(t_0)$$

The current through an inductor cannot change instantly.

流过电感器的电流直接正比于其两端电压对时间的积分：

$$i = \frac{1}{L}\int_{-\infty}^{t} v\, dt = \frac{1}{L}\int_{t_0}^{t} v\, dt + v(t_0)$$

流过电感器的电流不能突变。

5 Section Objectives：
(1) Understand the behaviors of capacitor and inductor in terms of its voltage and current.
(2) Learn about how to combine capacitors and inductors in series and parallel.

本节学习目标：
(1) 理解电容和电感的伏安特性；
(2) 了解电容和电感的串并联。

第七节　一阶电路和二阶电路的时域分析
Section 7　Time-domain Analysis of First-order and Second-order Circuits

RC（电阻—电容）电路
RC (resistor-capacitor) circuit
RL（电阻—电感）电路
RL (resistor-inductor) circuit
RLC 并联电路　parallel RLC circuit
RLC 串联电路　series RLC circuit
冲激函数　impulse function
冲激响应　impulse response
初始电压　initial voltage
初值　initial value
单位冲激函数　unit impulse function
单位阶跃函数　unit step function
断路器　circuit breaker
换路　switching
奇函数　odd function

阶跃函数　step function
阶跃响应　step response
介电常数　permittivity
绝缘体　insulator
开关　switch
开关函数　switching function
连续的　continuous
零输入响应　source-free response
零状态　zero state
零状态响应　zero-state response
偶函数　even function
奇异函数　singularity function
强迫响应　forced response
强制分量　forced component
取样性质　sampling property

全响应　complete response	一阶常微分方程式　first-order ordinary differential equation
筛分性质　sifting property	一阶电路　first-order circuit
时间常数　time constant	暂态/瞬变状态　transient state
衰减　decay	暂态分量　transient component
衰减常数　attenuation constant	暂态分析　transient analysis
特解　particular solution	暂态响应　transient response
特征方程　characteristic equation	指数规律衰减　exponential decay
特征根　characteristic root	滞后　delay/ lag
稳态响应　steady-state response	终值　final value
稳态值　steady-state value	自然响应　natural response
无源/零输入 RC 电路　source-free RC circuit	自由分量　free component
无源/零输入 RL 电路　source-free RL circuit	

1 The analysis in this section is applicable to any circuit that can be reduced to an equivalent circuit that is made up of a resistor and a single energy-storage element (inductor or capacitor). Such a circuit is first-order because its behavior is described by a first-order differential equation. When analyzing RC and RL circuits, one must always keep in mind that the capacitor is an open circuit to steady-state DC conditions while the inductor is a short circuit to steady-state DC conditions.

本节对电路的分析方法适用于任何一阶电路,即最终可简化为由一个电阻和一个储能元件(电感或电容)组成的等效电路的任何电路。一阶电路可以用一阶微分方程来描述。分析 RC 和 RL 电路时,不要忘记,在直流条件下,稳态下的电容器相当于开路,而电感器则相当于短路。

2 The time constant of a circuit is the time required for the response to decay by a factor of 1/e or 36.8 percent of its initial value. The time constant of an RL circuit equals the equivalent inductance divided by the Thévenin resistance as viewed from the terminals of the equivalent inductor. The time constant of an RC circuit equals the equivalent capacitance divided by the Thévenin resistance as viewed from the terminals of the equivalent capacitor.

电路的时间常数是电路相应衰减到初始值的 1/e 或 36.8% 时所需要的时间。RL 电路的时间常数等于等效电感除以等效电感两端看到的戴维南电阻。RC 电路的时间常数等于等效电容乘以等效电容两端看到的戴维南电阻。

3 Capacitive voltages and inductive currents are continuous; that is, they have the same value at $t = t_0^-$ and $t = t_0^+$. Where $t = t_0^-$ denotes the time just before a switching event, and $t = t_0^+$ is the time just after the switching event, assuming that the switching event takes place at t=0.

电容电压和电感电流是连续的,即它们在 $t = t_{0-}$ 和 $t = t_{0+}$ 时具有相同的值。式中 $t = t_{0-}$ 表示开关动作之前的瞬间, $t = t_{0+}$ 表示开关动作之后的瞬间,并假设开关动作时刻为 $t = 0$。

4 The natural response or transient response is the circuit's temporary response that will die out with time. The forced response or steady-state response is the behavior of the circuit a long time after an external excitation is applied. The total or complete response consists of the natural response and the forced response.

自然响应或暂态响应是电路的临时响应,它将随时间的延长而消失。强迫响应或稳态响应是加上外部激励后的电路长时间的行为特征。电路的完全响应由其暂态响应和稳态响应组成。

5 The natural response is obtained when no independent source is present. It has the general form

$$x(t) = x(0)e^{-t/\tau}$$

where x represents current through (or voltage across) a resistor, a capacitor, or an inductor, and $x(0)$ is the initial value of x.

对于没有独立电源的电路,其响应为自然响应(或暂态响应),它的一般形式为:

$$x(t) = x(0)e^{-t/\tau}$$

这里 x 表示流过电阻、电容(或电感)的电流,或者表示电阻、电容(或电感)两端的电压。$x(0)$ 是 x

第五章 电路原理

的初始值。

6 The step response is the response of the circuit to a sudden application of a dc current or voltage. Finding the step response of a first-order circuit requires the initial value $x(t_0^+)$, the final value $x(\infty)$, and the time constant τ. With these three items, we obtain the step response as

$$x(t) = x(\infty) + [x(t_0^+) - x(\infty)]e^{-t/\tau}$$

A more general form of this equation is

$$x(t) = x(\infty) + [x(t_0^+) - x(\infty)]e^{-(t-t_0)/\tau}$$

Or we may write it as

Instantaneous value = Final + [Initial − Final] $e^{-(t-t_0)/\tau}$

阶跃响应是电路突然加上一个直流电流或电压时的电路响应。得到一阶电路的阶跃响应需要知道初始值、终止值和时间常数。有了这三个值,则阶跃响应是:

$$x(t) = x(\infty) + [x(t_0^+) - x(\infty)]e^{-t/\tau}$$

更一般的表达式为:

$$x(t) = x(\infty) + [x(t_0^+) - x(\infty)]e^{-(t-t_0)/\tau}$$

或可写成:

瞬时值＝终值＋[初值-终值] $e^{-(t-t_0)/\tau}$

7 Section Objectives：
(1) Understand the basic concepts of first-order circuit.
(2) Learn how to use the three elements method to solve the first-order circuit.
(3) Understand source-free RC and RL circuits.
(4) Learn about step and impulse response RC and RL circuits.

本节学习目标:
(1) 理解一阶电路的基本概念;
(2) 学会使用三要素法求解一阶电路;
(3) 理解 RC 和 RL 电路的零输入响应;
(4) 了解 RC 和 RL 电路的阶跃和冲激响应。

第八节 相量法
Section 8 Phasor Method

参考相量　reference phasor	频率　frequency
超前　advance/ lead	时域　time domain
初相　initial phase	实部　real part
电流相量　current phasor	同相　in phase
电路元件方程的相量形式　phasor relations for circuit elements	相角　phase angle
	相量　phasor
电压相量　voltage phasor	相量法　phasor method
反相　opposite phase	相量图　phasor diagram
幅频特性　amplitude-frequency characteristic	相频特性　phase-frequency characteristic
幅值/振幅　amplitude	相位差　phase difference
幅值相量　amplitude phasor	虚部　imaginary part
复平面　complex-number plane	异相　out of phase
极坐标形式　polar form	正弦函数　sinusoidal function
角频率　angular frequency	直角坐标形式　rectangular form
欧拉公式　Euler's identity	指数形式　exponential form

1 A sinusoid is a signal that has the form of the sine or cosine function.
正弦量是有正弦或余弦函数形式的信号。

2 A phasor is a complex number that represents the amplitude and phase of a sinusoid. Given the sinusoid $u(t) = U_m \cos(\omega t + \varphi)$, its phasor $\dot{U}$ is
$$\dot{U} = U_m \angle \varphi$$
相量是一个复数,能表示正弦量的幅值和相位。给定正弦量 $u(t) = U_m \cos(\omega t + \varphi)$,其相量 $\dot{U}$ 是:
$$\dot{U} = U_m \angle \varphi$$

3 In AC circuits, voltage and current phasors always have a fixed relation to one another at any moment of time. If $u(t) = U_m \cos(\omega t + \varphi_u)$ represents the voltage through an element and $i(t) = I_m \cos(\omega t + \varphi_i)$ represents the current through an element, then $\varphi_i = \varphi_u$ if the element is a resistor, φ_i leads φ_u by 90° if the element is a capacitor, and φ_i lags φ_u by 90° if the element is an inductor.

在交流电路中,电压和电流的相量在任何时刻(瞬间)都有将它们联系起来的固定关系。若 $u(t) = U_m \cos(\omega t + \varphi_u)$ 表示元件两端的电压, $i(t) = I_m \cos(\omega t + \varphi_i)$ 表示流过该元件的电流,则:若该元件是电阻,那么 $\varphi_i = \varphi_u$;若是电容器,那么 φ_i 超前于 φ_u 90°;若是电感器,那么 φ_i 滞后于 φ_u 90°。

4 Basic circuit laws (Ohm's and Kirchhoff's) apply to AC circuits in the same manner as they do for DC circuits.

电路的基本定律(欧姆定律和基尔霍夫定律)也适用于交流电路,其形式与直流电路中的基本定律一样。

5 The differences between $u(t)$ and $\dot{U}$ should be emphasized:
(1) $u(t)$ is the instantaneous or time-domain representation, while $\dot{U}$ is the frequency or phasor-domain representation.
(2) $u(t)$ is time dependent, while $\dot{U}$ is not. (This fact is often forgotten by students.)
(3) $u(t)$ is always real with no complex term, while $\dot{U}$ is generally complex.

注意以下几点 $u(t)$ 与 $\dot{U}$ 的差别:
(1) $u(t)$ 是瞬时或时域的表示,而 $\dot{U}$ 是频率或相量域的表示。
(2) $u(t)$ 是与时间有关的,而 $\dot{U}$ 不与时间有关(这一点学生常常不注意)。
(3) $u(t)$ 总是实数且没有复数项,而 $\dot{U}$ 一般都为复数。

6 Section Objectives:
(1) Learn about the concepts of sinusoids and phasors.
(2) Apply phasors to circuit elements.

本节学习目标:
(1) 了解正弦量和相量的基本概念;
(2) 掌握电路元件的相量表示。

第九节 正弦稳态电路的分析
Section 9　Sinusoidal Steady-state Analysis

波形	waveform	电纳	susceptance
代数形式	algebraic form	电位计	potentiometer
导纳	admittance	电压符号	voltage symbol
导纳三角形	admittance triangle	电压极性	voltage polarity
等效变换	equivalent transformation	电阻性阻抗	resistive impedance
等效导纳	equivalent admittance	端点电压	terminal voltage
等效电抗	equivalent reactance	额定电流	rated current
等效电纳	equivalent susceptance	额定电压	rated voltage
等效阻抗	equivalent impedance	额定功率	rated power
电度表	watthour meter	额定容量	rated capacity
电抗	reactance	峰值	peak value
电流方向	current direction	复功率	complex power

第五章 电路原理

复阻抗	complex impedance	实际直流电流源	practical DC current source
感抗	inductive reactance	实际直流电压源	practical DC voltage source
感纳	inductive susceptance	视在功率	apparent power
感性	inductive	顺时针方向	clockwise
感性阻抗	inductive impedance	瞬时功率	instantaneous power
功率三角形	power triangle	瓦特表	wattmeter
功率因数	power factor	无功分量	reactive component
功率因数的提高	power factor correction	无功伏安/乏	reactive volt-ampere(var)
功率因数角	power factor angle	无功功率	reactive power
可变电阻	variable resistor	有功分量	active component
能量守恒定律	law of conservation of energy	有功功率	real power
逆时针方向	anti-clockwise	正弦量	sinusoid variable
欧姆表	Ohmmeter	正弦稳态	sinusoidal steady state
平均功率	average power	正弦稳态响应	sinusoidal steady-state response
平均值	average value	直流电阻	DC resistance
千瓦时	kilowatthours(kWh)	阻抗	impedance
容抗	capacitive reactance	阻抗的并联	impedances in parallel
容纳	capacitive susceptance	阻抗的串联	impedances in series
容性	capacitive	阻抗模	impedance module
容性阻抗	capacitive impedance	阻抗三角形	impedance triangle

1 The impedance Z of a circuit is the ratio of phasor voltage across it to the phasor current through it. The admittance Y is the reciprocal of impedance. Impedances are combined in series or in parallel the same way as resistances in series and in parallel; that is, impedances in series add while admittances in parallel add.

电路的阻抗 Z 是电路两端的电压相量与流过它的电流相量之比,导纳 Y 是阻抗的倒数。阻抗的串联和并联与电阻的串并联方法相同,即串联时阻抗相加,并联时导纳相加。

2 We apply nodal and mesh analysis to AC circuits by applying KCL and KVL to the phasor form of the circuits.

由于 KCL 和 KVL 适用于电路的相量形式,所以,可以用结点电压法和网孔电流法等方法来分析交流电路。

3 Steps to analyze AC circuits:
(1) Transform the circuit to the phasor or frequency domain;
(2) Solve the problem using circuit techniques (nodal analysis, mesh analysis, superposition, etc.).
(3) Transform the resulting phasor to the time domain.

分析交流电路的步骤:
(1) 将电路转换到频域或相量域;
(2) 用各种方法(结点分析、网孔分析、叠加原理等)求解电路;
(3) 将频域的结果转换到时域。

4 The average power is the average of the instantaneous power over one period. The apparent power (in VA) is the product of the rms values of voltage and current. A resistive load (R) absorbs power at all times, while a reactive load (L or C) absorbs zero average power.

平均功率是一个周期内瞬时功率的平均值。视在功率(VA)是电压和电流有效值的乘积。电阻负载(R)任何时候都吸收功率,而电抗负载吸收的平均功率为零。

5 For maximum average power transfer, the load impedance must be equal to the complex conjugate of the Thévenin impedance.

最大平均功率传输条件是负载阻抗必须等于戴维南阻抗的共轭复数。

6 The principle of conservation of AC power states that the complex, real and reactive powers of the sources equal the respective sums of the complex, real and reactive powers of the individual loads.

交流功率的守恒定理：电源的复功率、有功功率、无功功率等于它在各个单独负载上的复功率、有功功率、无功功率之和。

7 The power factor is the cosine of the phase difference between voltage and current:
$$\lambda = pf = \cos(\theta_v - \theta_i)$$

It is also the cosine of the angle of the load impedance or the ratio of real power to apparent power. The pf is lagging if the current lags voltage (inductive load) and is leading when the current leads voltage (capacitive load).

功率因数是电压和电流相位差的余弦函数：
$$\lambda = pf = \cos(\theta_v - \theta_i)$$

功率因数也是负载阻抗角的余弦函数，或者是有功功率与视在功率之比。若电流滞后于电压（电感性负载），则 pf 是滞后的；若电流超前于电压（电容性负载），则 pf 是超前的。

8 The process of increasing the power factor without altering the voltage or current to the original load is known as power factor correction. Power factor correction is necessary for economic reasons. It is the process of improving the power factor of a load by reducing the overall reactive power.

不改变原始负载的电压、电流而提高功率因数的过程称为功率因数的改进。从经济原因考虑，功率因数的提高是必需的。改进负载功率因数也就是降低了总的无功功率。

9 Section Objectives:
(1) Learn how to apply previously learned circuit techniques to sinusoidal steady-state analysis.
(2) Understand the concepts of AC powers.
(3) Learn about power factor correction.
(4) Understand the principle of conservation of AC power.

本节学习目标：
(1) 学会如何将之前所学的电路分析方法用于正弦稳态电路分析；
(2) 理解交流电路功率的概念；
(3) 会功率因数修正；
(4) 理解交流功率守恒原理。

第十节 含有耦合电感的电路
Section 10　Magnetically Coupled Circuits

T形网络　T-connected network
变比　transformation ratio
变压器　transformer
初级电路/原边　primary circuit
初级线圈　primary winding
磁场　magnetic field
磁链　flux linkage
磁耦合　magnetically couple
磁通　magnetic flux
次级电路/副边　secondary circuit
次级线圈　secondary winding
电感的串并联　series-parallel combinations of inductance
电容串并联　series-parallel combinations of capacitance
二端口耦合电感元件　two-port coupled inductors
反接/反向串联　inversed connection/connection in opposition
符号　notation
互感　mutual inductance
互感电抗　mutual inductive reactance
空心变压器　air-core transformer
空心线圈　air-core coil
理想变压器　ideal transformer
耦合电感元件　coupled inductor
耦合系数　coefficient of coupling
耦合线圈　coupled coil
顺接/正向串联　connection in aiding
四端元件　four-terminal component

第五章 电路原理

铁芯线圈　iron-core coil
同名端　dotted terminals
韦伯　weber (Wb)
线圈匝数　number of turns on the coil
匝数比　turns ratio
自感　self-inductance

1 The polarity of the mutually induced voltage is expressed in the schematic by the dot convention. The dot convention is stated as follows: If a current enters the dotted terminal of one coil, the reference polarity of the mutual voltage in the second coil is positive at the dotted terminal of the second coil. Alternatively, if a current leaves the dotted terminal of one coil, the reference polarity of the mutual voltage in the second coil is negative at the dotted terminal of the second coil.

互感电感的极性在电路中的表达要遵守同名端规则。同名端规则是：若电流进入一个线圈的同名端，则在第二个线圈的同名端处，其互感电压的参考极性是正的。或者：若电流离开一个线圈的同名端，则在第二个线圈的同名端处，其互感电压的参考极性是负的。

2 An ideal transformer is a unity-coupled, lossless transformer in which the primary and secondary coils have infinite self-inductances. A step-down transformer is one whose secondary voltage is less than its primary voltage. A step-up transformer is one whose secondary voltage is greater than its primary voltage.

理想变压器是耦合系数为1的、无损耗的而且初级和次级线圈有无穷大自感量的变压器。降压变压器的次级电压小于其初级电压，升压变压器的次级电压大于其初级电压。

3 Section Objectives:
(1) Understand magnetically coupled circuits.
(2) Learn the concept of mutual inductance.
(3) Learn how to analyze circuits involving mutual inductance.

本节学习目标：
(1) 理解含有磁耦合的电路；
(2) 了解互感的概念；
(3) 掌握含有互感的电路的分析方法。

第十一节　电路的频率响应
Section 11　Frequency Response of Circuit

并联谐振　parallel resonance
并联谐振阻抗　parallel resonance impedance
传递函数　transfer function
串联谐振　series resonance
串联谐振阻抗　series resonance impedance
带宽　bandwidth
带通滤波器　band-pass filter
带阻滤波器　band-reject filter
低通滤波器　low-pass filter
电路选择性　selectivity of the circuit
动态电路　dynamic circuit
动态电阻　dynamic resistance
动态线路　dynamic route
高通滤波器　high-pass filter
共轭匹配　conjugate matching
固定电阻　fixed resistor
固有频率/自然频率　natural frequency
计算机辅助电路分析　computer-aided circuit analysis
频带　frequency band
频率响应　frequency response
频率选择电路　frequency selective circuit
品质因数　quality factor
网络函数　network function
无源滤波器　passive filter
无源线性电路　passive linear circuit
谐振电路　resonant circuit
谐振电路阻抗　resonant circuit impedance
谐振频率　resonant frequency
谐振阻抗　resonance impedance
选频特性　frequency-selection characteristic

1 Resonance is a condition in an RLC circuit in which the capacitive and inductive reactances are equal in magnitude, thereby resulting in a purely resistive impedance.

谐振是RLC电路中的一种状态，该电路中电容和电感的电抗大小是相等的，结果呈现出纯电阻的阻抗性质。

2 If we let the amplitude of the sinusoidal source remain constant and vary the frequency, we obtain the circuit's frequency response.

如果我们保持正弦交流电源的幅值不变而改变它的频率就可以得到电路的频率响应。

3 Section Objectives：
(1) Understand the concept of transfer function.
(2) Learn about series and parallel resonant RLC circuits.

本节学习目标：
(1) 理解传递函数的概念；
(2) 了解 RLC 串联和并联谐振电路。

第十二节　三相电路
Section 12　Three-phase Circuit

ABC/或正/相序	ABC /or positive/ sequence
ACB/或负/相序	ACB /or negative/ sequence
A 相电压	A-phase voltage
B 相电压	B-phase voltage
C 相电压	C-phase voltage
Y-Y 电路	wye-wye(Y-Y) circuit
不对称三相电路	unbalanced three-phase circuit
传输线阻抗	transmission-line impedance
单相等效电路	single-phase equivalent circuit
对称三相电路	balanced three-phase circuit
两表法	two-wattmeter method
三表法	three-wattmeter method
三相电路	three-phase circuit
三相电源	three-phase voltage source
三相三线制	three-phase three-wire system
三相四线制	three-phase four-wire system
三相制	three-phase system
瞬时电压	instantaneous voltage
线电流	line current
线电压	line voltage
相电流	phase current
相电压	phase voltage
中点	neutral terminal
中线	neutral wire

1 The phase sequence is the order in which the phase voltage of a three-phase generator occurs with respect to time. In an ABC sequence of balanced source voltages, $\dot{U}_{an}$ leads $\dot{U}_{bn}$ by 120°, which in turn leads $\dot{U}_{cn}$ by 120°. In an ACB sequence of balanced voltages, $\dot{U}_{an}$ leads $\dot{U}_{cn}$ by 120°, which in turn leads $\dot{U}_{bn}$ by 120°.

相序是三相发电机相电压产生的时间顺序。对称电压源的 ABC 相序情况是 $\dot{U}_{an}$ 超前于 $\dot{U}_{bn}$ 120°，$\dot{U}_{bn}$ 又超前于 $\dot{U}_{cn}$ 120°。对称电源的 ACB 相序是 $\dot{U}_{an}$ 超前于 $\dot{U}_{cn}$ 120°，$\dot{U}_{cn}$ 又超前于 $\dot{U}_{bn}$ 120°。

2 The easiest way to analyze a balanced three-phase circuit is to transform both the source and the load a Y-Y system and then analyze the single-phase equivalent circuit. An unbalanced three-phase system can be analyzed using nodal or mesh analysis.

分析对称三相电路的最方便的方法是将电源和负载都转换成 Y-Y 系统，然后，再分析其单相等效电路。对于不对称三相系统的分析可以用结点电压法或网孔分析法。

3 The total instantaneous power in a balanced three-phase system is constant and equal to the average power.

对称三相系统的总瞬时功率是个常数，且等于其平均功率。

4 The total real power is measured in three-phase systems using either the three-wattmeter method or the two-wattmeter method.

三相系统总有功功率的测量可以用三表法，也可以用两表法。

5 Section Objectives：
(1) Be familiar with the concepts of three-phase circuits.
(2) Understand how to analyze balanced three-phase circuits.
(3) Learn about power in a balanced three-phase system.

(4) Know how to analyze unbalanced three-phase systems.

本节学习目标：
(1) 掌握三相电路基本概念；
(2) 理解对称三相电路的分析；
(3) 了解对称三相电路中的功率；
(4) 了解不对称三相电路的分析。

第十三节　拉普拉斯变换
Section 13　The Laplace Transform

s 域的等效电路　s-domain equivalent
部分式法/分解定理　partial fraction expansion method
初始电流　initial current
初始条件　initial condition
初值定理　initial-value theorem
单极点　simple pole
分部积分法　integration by part
附加电流源　appended current source
附加电压源　appended voltage source
傅里叶变换　Fourier transform
积分性质　time integration property
极点　pole
假分式　improper rational function
拉普拉斯变换　Laplace transform
拉普拉斯变换对　Laplace transform pairs
拉普拉斯反变换　inverse Laplace transform
零初始条件　zero initial condition
零点　zero
留数法　residue method
频域函数　phasor domain function
时域函数　time domain function
实数单根　real simple root/distinct real root
收敛因子　convergence factor
完全配方法　completing the square
微分性质　time differentiation property
线性性质　linearity property
延迟性质　time-delay property/ time-shift property
一阶极点　first-order pole
有理真分式　proper rational function
原函数　original function
运算导纳　operation admittance
运算电路　operation circuit
运算法　operation method
运算阻抗　operation impedance
展开系数　expansion coefficient
指数函数　exponential function
终值定理　final-value theorem
重极点　repeated pole

1 Steps in applying the Laplace transform:
(1) Transform the circuit from the time domain to the s-domain.
(2) Solve the circuit using nodal analysis, mesh analysis, source transformation, superposition, or any circuit analysis technique with which we are familiar.
(3) Take the inverse Laplace transform of the solution and thus obtain the solution in the time domain.

用拉普拉斯变换法分析电路的步骤：
(1) 将电路从时域转换为 s 域的等效电路；
(2) 用结点电压法、网孔电流法、电源变换原理、叠加原理以及所学过的其他电路分析技术求解电路；
(3) 对得到的电路解进行拉普拉斯逆变换，最后得到电路的时域解。

2 Section Objectives:
(1) Become familiar with the concepts of Laplace transform.
(2) Learn how to analyze linear circuits with Laplace transform.

本节学习目标：
(1) 掌握拉普拉斯变换的基本概念；
(2) 学会用拉普拉斯变换分析线性电路。

第十四节　二端口网络
Section 14　Two-port Networks

导抗　immittance
导纳参数　admittance parameter
导纳参数矩阵　admittance parameter
短路导纳参数矩阵　short-circuit admittance parameter
短路输出导纳　short-circuit output admittance
短路输入导纳　short-circuit input admittance
短路输入阻抗　short-circuit input impedance
短路转移导纳　short-circuit transfer admittance
对称的　symmetrical
二端口网络　two-port networks
二端口元件　two-port element
开路输出导纳　open-circuit output admittance
开路输出阻抗　open-circuit output impedance
开路输入阻抗　open-circuit input impedance
开路转移阻抗　open-circuit transfer impedance
开路阻抗参数矩阵　open-circuit impedance parameter
驱动点导纳　driving-point admittance
驱动点阻抗　driving-point impedance
线性网络　linear network
一端口网络　one-port networks
转移导纳　transfer admittance
转移函数　transfer function
转移阻抗　transfer impedance
阻抗参数　impedance parameter
阻抗参数矩阵　impedance parameter

1 A two-port network is an electrical network with two separate ports for input and output. The six parameters used to model a two-port network are the impedance [z], admittance [y], hybrid [h], inverse hybrid [g], transmission [T], and inverse transmission [t] parameters. The parameters can be calculated or measured by short-circuiting or open-circuiting the appropriate input or output port.
二端口网络是一个有两个分开端口作为输入和输出的电子网络。有六个参数可描述双端口的性质：阻抗[z]、导纳[y]、混合[h]、逆混合[g]、传输[T]和逆传输[t]参数。双端口参数可以由短路或开路相应的输入或输出端口来计算或测量。

2 Section Objectives：
(1) Understand the concepts of two-port networks.
(2) Understand several parameters of two-port networks.
本节学习目标：
(1) 理解二端口网络的概念；
(2) 理解二端口网络的几种参数矩阵。

第六章 电气控制技术与 PLC 应用
Chapter 6 Electrical Control and Programmable Controller

第一节 电气控制技术
Section 1 Electrical Control

按钮　push button
保险丝熔断　fuse blown
常闭触点　normally closed contact
传感器信号　sensor signal
电磁阀　solenoid valve
短路保护　short-circuit protection
额定功率　rated power
负载电路　load circuit
工作原理图　function diagram
过流保护　overcurrent protection
互锁　interlock
急停按钮　emergency stop push button
继电器触点　relay contact
接线端块　fan-out connector
开关柜　control cabinet
漏电流　leakage current
灭弧元件　quenching element
设备名称　device name
时间继电器　timing relays
限位开关　limit switches
仪器和控制　instrumentation and control
直流电源　DC power supply
主控继电器　Master Control Relay
自动化系统　automation system

The electrical section contains the solid-state proportional temperature controllers and solenoid valves.

电子控制室内有固体线路的比例温度控制器和电磁阀。

第二节 可编程控制器概述
Section 2 Introduction to Programmable Controller

DP 主站系统　DP master system
编程语言　Programming languages
常闭　normally closed
程序执行　program execution
处理阶段　processing phase
电源故障　power failure
动作时间　action time
分布式 I/O 设备　distributed I/O device
工作存储器　work memory
功能块　functional blocks
过程映像输入寄存器
process-image input register location
监视器　load monitor
结构化文本　structured text
金属导轨　metal guide
可编程控制器　programmable controller
可编程逻辑控制　programmable logic control
扫描周期　scan cycle
适配器电缆　adapter cable
手持编程器　handheld programming unit
输入输出接口　Input/output interface
梯形图　ladder diagrams
停止系统模式　stop system mode
虚拟背板总线　virtual backplane bus
应用程序存储器　application memory
硬件接线　hard-wired
远程 I/O 子系统　remote I/O subsystem
指令表　instruction lists
组态选项　Configuration Options

1 Programmable logic controllers, also called programmable controllers or PLCs, are solid-state members of the computer family, using integrated circuits in stead of electromechanical devices to implement control functions.

可编程逻辑控制器,也叫作可编程控制器或 PLC,是计算机家庭的重要成员。它使用集成电路代替机电设备去实现控制功能。

2 The first PLCs offered relay functionality, thus replacing the original hardwired relay logic, which used electrically operated devices to mechanically switch electrical circuits. They met the requirements of modularity, expandability, programmability, and ease of use in an industrial environment. These controllers were easily installed, used less space, and were reusable.

第一台 PLC 提供了继电器控制功能，从而替换原有的使用硬接线的逻辑继电器控制电路。它们满足模块化、可扩展性、可编程性和用于工业环境的要求。这些控制器安装容易，占用空间较少，而且可重复使用。

3 A PLC works by continually scanning a program. We can think that this scan cycle consists of three important steps - CHECK INPUT STATUS, EXECUTE PROGRAM-NEXT, and UPDATE OUTPUT STATUS - to represent events. After the third step the PLC goes back to step one and repeats the steps continuously. One scan time is defined as the time it takes to execute the three steps listed above.

PLC 是通过连续扫描一个程序来工作的。我们可以认为扫描周期是由三个主要阶段组成：检测输入状态、执行程序和更新输出状态。PLC 执行完第三阶段后就返回第一阶段，并反复循环。一次扫描时间定义为执行上面三个阶段所花的时间。

第三节　组件与系统
Section 3　Components and Systems

4 线制传感器　4-wire transducer
DIN 导轨　DIN rail
L 端子　L terminal
主站接口模块　master interface module
信号模块　signal module
光隔离　optically isolated
分布式 I/O 设备　distributed I/O device
前连接器模块　front connector module
占位模块　dummy module
地址分配列表　address assignment list
地址标识符　address identifier
备用电池　back-up battery
外设输出双字　peripheral output double word
寻址错误　addressing error
底座连接器　base connector
总电流　aggregate current
扩展模块　extension module

接地故障检测器　ground fault detector
数字量输入/输出模块　digital input/output module
晶体管输出　transistor output
模块背板　module backplane
模拟器模块　simulator module
模拟量信号　analog signal
模拟量输入模块　analog input module
模拟量输出模块　analog output module
现场总线隔离变压器　field bus isolating transformer
电源模块　power supply module
直接 I/O 访问　direct I/O access
硬件组态　hardware configuration
背板总线　backplane bus
订货号　order number
输出电路　output circuit
附加端子　add-on terminal

1 The PLC system is chassis-based and provides the option to configure a control system that uses sequential, process, motion, and drive control in addition to communication and I/O capabilities. This section describes some of the many system configuration options that are available with controllers.

PLC 系统是基于机架的系统，对使用顺序控制、过程控制、运动控制和驱动控制的控制系统可以提供进行组态的选项，还可提供通信和 I/O 功能。在一个机架中装配 I/O 的独立控制器是其中一种最简单的组态。

2 When you design a system, there are several system components to consider for your application. Some of these components include all of the following:
- I/O devices
- Motion control and drive requirements

- Communication modules
- Controllers
- Chassis
- Power supplies
- Software

设计时，需要根据您的应用而考虑多种系统组件。其中一些组件包括以下部分：
- I/O 设备
- 运动控制和变频器要求
- 通信模块
- 控制器
- 机架
- 电源
- 软件

3 You can fill the slots of your chassis with any combination of controllers, communication modules, and I/O modules.

可以将控制器、通信模块和 I/O 模块以任意组合方式插入机架的插槽中。

第四节　PLC 编程
Section 4　PLC PROGRAMMING

DB 编辑器　DB editor	indirect addressing
RLO 沿检测　RLO edge detection	块调用命令　block call command
SCL 编译器　SCL compiler	块接口　block interface
背景数据块　instance data block	块结构　block architecture
变量声明表　variable declaration table	块输入　block input
程序/流程控制指令　program/flow control instructions	块图标　Block icon
触点　contact	立即寻址　immediate addressing
存储状态字的溢出位　stored overflow bit of the status word	逻辑操作结果　result of logic operation
错误组织块　error organization block	内置指针　built-in pointer
定时器类型　type of timer	暖启动　warm restart
二进制结果位　binary result bit	强制值　force value
符号编辑器　symbol editor	任意跳转　any transition
负跳沿跳转　negative edge transition	时钟存储器　clock memory
功能块库　function block library	使能输入/输出　enable input / out put
共享数据块　shared data block	输入/输出参数　in/out parameter
基本项目　basic project	数据操作指令　data manipulation instructions
计数范围　counting range	数据块　data block
计数器和比较指令　counter and comparison instruction	数据块寄存器　data block register
寄存器间接寻址　register indirect addressing	数据类型声明　data type declaration
接通延迟定时器　on-delay timer	数据字节 DBB　data byte DBB
静态变量　static variable	数组元素　array element
绝对地址　absolute address	双精度整数　Double Integer
绝对寻址　absolute addressing	算数运算指令　arithmetic instructions
跨区域寄存器间接寻址　area-crossing register-	梯形图编辑器　Ladder Editor
	跳转目标　jump destination
	网络通信指令　network communications
	位存储器地址区　bit memory address area

位累加器 bit accumulator	用一条语句进行的 DB 访问 DB access with one statement
位指令 bit instruction	优先级等级 priority class
线圈 coil	语句表 statement list
行注释 line comment	中断堆栈 interrupt stack
循环程序处理 cyclic program processing	中断组织块 interrupt organization block
循环控制 Loop control	装载用户程序 load user program
沿检测 edge detection	
移位和循环指令 shift and rotate instructions	

1 The original ladder diagrams were established to represent hardwired logic circuits used to control machines or equipment. Due to wide industry use, they became a standard way of communicating control information from the designers to the users of equipment.

最初，建立阶梯图是为了替代硬逻辑电路用于机器或设备的控制。由于广泛的工业应用，它们成为从设计师到用户之间传递控制信息的一个途径。

2 Instruction: You organize ladder diagram as rungs on a ladder and put instructions on each rung. There are two basic types of instructions: Input instruction: An instruction that checks, compares, or examines specific conditions in your machine or process. Output instruction: An instruction that takes some action, such as turn on a device, turn off a device, copy data, or calculate a value.

指令：可在梯形图程序上设置多行来构建梯形图，并在每个行上添加指令。
有两种基本指令类型：输入指令：检查、对比或检验机器或进程中的指定条件的指令。输出指令：采取某些操作的指令，如打开设备、关闭设备、复制数据或计算值。

3 Function block diagram (FBD) is a graphical language that allows the user to program elements (e. g., PLC function blocks) in such a way that they appear to be wired together like electrical circuits.

功能框图(FBD)是一种图形化的语言，用户程序元素(例如，PLC 的功能模块)以类似电路的方式连接在一起。

第五节 顺序控制系统
Section 5　Sequence Control System

备择路径 alternative path	分支和合并 branch & merge
步骤启用 step enabling	顺序功能图 sequential function chart
动作 action	转换 transitions
分支 branch	

1 Grafcet (Graphe Fonctionnel de Commande étape Transition) is a symbolic, graphic language, which originated in France, which represents the control program as steps or stages in the machine or process. In fact, the English translation of Grafcet means "step transition function charts."

Grafcet(Graphe Fonctionnel de Commande étape Transition)是一个象征性的图形语言，它起源于法国。它用控制程序来描述机器或过程的步骤或阶段。事实上，顺序控制的英语翻译是"步转换函数图表"。

2 The SFC programming framework contains three main elements that organize the control program:
- steps
- transitions
- actions

A step is a stage in the control process. An action is a set of control instructions prompting the PLC to execute a certain control function during that step. An action may be programmed using any one of the four IEC 1131-3 languages. After the PLC executes a step/action, it must receive a transition before it will proceed to the next step. A transition can take the form of a variable input, a result of a previous action, or a conditional IF statement.

第六章 电气控制技术与 PLC 应用

顺序功能图编程框架包含三个主要的部分,这三部分组成了控制程序:
- 步
- 转换
- 动作

步骤是控制过程中的一个阶段。动作是一组控制指令,使 PLC 执行当前步的一定的控制功能。一个行动可使用四种 IEC 1131-3 语言中的任何一种编程。在 PLC 执行一个步骤/行动以后,它必须满足一个转换条件,才能继续下一步。转换条件可以使用一个变量输入、前面的操作结果或者一个条件型 IF 语句。

第六节 通信与网络
Section 6 Communication and Network

DP 标识符　　DP identifier
DP 标准从站　　DP standard slave
DP 从站接口　　DP slave interface
DP 延迟时间　　DP delay time
I/O 总线网络　　I/O bus network
MPI 卡　　MPI card
PROFIBUS DP 主站　　PROFIBUS DP master
PROFIBUS 总线连接器　　PROFIBUS bus connector
传输控制协议/网际协议　　transmission control protocol/internet protocol (TCP/IP)
串行端口通信　　Serial port communication

光纤电缆　　fiber-optic cable
介质访问控制　　medium access control (MAC)
局域网　　local area network (LAN)
内置通信端口　　Communication ports
通讯功能块　　communication function block
网络组态　　network configuration
响应时间　　response time
总线节点　　bus node
总线连接　　bus connection
总线协议　　bus protocol

1 For some applications, a variety of devices may be connected to the chassis via multiple communication networks. For example, a system might be connected to the following:
- Distributed I/O via an Ethernet network
- A PowerFlex drive connected via a DeviceNet network
- Flowmeters connected via a HART connection

在某些应用中,可以通过多种通信网络将多种设备连接到机架。例如,系统可以:
- 通过以太网网络连接到分布式 I/O
- 通过 DeviceNet 网络连接到 PowerFlex 变频器
- 通过 HART 连接而连接到流量计

2 You can configure a percentage of the scan cycle to be dedicated for processing a run mode edit compilation or execution status. (Run mode edit and execution status are options provided by STEP 7—Micro/WIN to make debugging your program easier.) As you increase the percentage of time that is dedicated to these two tasks, you increase the scan time, which makes your control process run more slowly.

您可以设定一个扫描周期的百分比用来处理运行模式编辑或执行状态相关的通讯请求。(运行模式编辑和执行状态是 STEP 7—Micro/WIN 提供的备选功能,能使您更轻松地调试程序。)在您增加用于通讯请求处理时间百分比的同时,扫描时间也会随之增加,从而会导致控制过程运行速度变慢。

第七节 闭环控制系统
Section 7 Closed Loop Control Systems

比例积分控制器　　proportional-integral controller
比例控制器　　proportional controller
串级控制　　cascade control
防爆保护模拟量输出　　Ex analog output

防爆保护模拟量输入　　Ex analog input
连续控制　　continuous-mode controller
模块化闭环控制　　modular closed-loop control
模拟量信号　　analog signal

Most PLC applications that implement PID control employ automatic/manual control stations that allow the operator to switch between manual and PLC process control. To prevent a step change or "bump" during this switch, the control station must ensure that both controllers, the manual controller and the PLC (automatic), send the same output to the process. Otherwise, the process may receive a change in the control variable, which could produce a transient response in the system.

大多数采用自动/手动控制实现PID控制的PLC应用程序,都允许操作人员进行手动和PLC程序控制之间切换。为了防止在切换时的阶跃变化或"突变",控制系统必须确保两个控制器,手动控制器和PLC(自动),送出相同的输出。否则,控制系统可能因控制变量的改变而产生瞬态响应。

第七章　电气设计 CAD 软件
Chapter 7　Electrical Design Software

第一节　项目
Section 1　Projects

帮助系统	help system	三维建模	3D modeling
创建新图形	create new drawing	下拉菜单	pull-down menu
对话框	dialog box	线参考	line reference
二维草图与注释	2D drafting & annotation	项目管理器	project manager
工具栏	toolbar	项目设置	project setting
后缀设置	suffix setup	项目特性	project properties
基于参考的标记	reference-based tagging	选择模板	select template
交互参考	cross referencing	元件标记格式	component tag format

1 AutoCAD Electrical is a project-based system. An ASCII text with a .wdp extension defines each project. This project file contains a list of project information, default project settings, drawing properties, and drawing file names.

AutoCAD Electrical 是一个基于项目的系统。扩展名为 .wdp 的 ASCII 文本可定义每个项目。此项目文件包含项目信息、默认项目设置、图形特性和图形文件名的列表。

2 You can have an unlimited number of projects; however, only one project can be active at a time.

您可以拥有任意数量的项目，但每次只能激活一个项目。

3 A single project file can have drawings located in many different directories. There is no limit to the number of drawings in a project. When you create a drawing, using the new drawing tool, it is automatically added to the active project.

一个项目文件可以包含许多不同目录中的图形。一个项目中可以有任意数量的图形。当使用新建项目工具创建图形时，该图形将自动添加到激活项目中。

第二节　导线
Section 2　Wires

插入导线	inserting wire	删除线号	delete wire number
插入线号	insert wire number	水平导线	horizontal wire
垂直母线	vertical bus wire	添加横档	add rung
导线编号	wire numbering	线号	wire number
导线段	wire segment	信号箭头	signal arrow
导线图层	wire layer	修改阶梯	modify ladder
接线端子	wire connection terminal	修剪导线	trim wire
接线连接点	wire connection point	源箭头	source arrow
目标箭头	destination arrow	源信号	source signal
目标信号	destination signal		

1 AutoCAD Electrical considers two wire segments connect if the end of one wire segment touches or falls within a small trap distance of any part of the other wire segment. AutoCAD Electrical considers a wire connected to a component if the wire end falls within a trap distance from the wire connection-point attribute of a component.

如果一条导线段的末端接触或落在另一条导线任何部分的较小捕获范围内，则 AutoCAD Electrical 认为这两条导线段相连。如果导线的末端落在元件接线点属性的捕获范围内，则 AutoCAD

Electrical 认为导线和元件相连。

2 You can start or end a wire segment in empty space, from an existing wire segment, or from an existing component.
可以在空白区域、从现有导线段或从现有元件中开始或结束另一个导线段。

3 Wire numbers can be assigned to any existing wires on an individual selection, an entire drawing, selected drawings in a project, or an entire project.
可以将线号指定给单独选择的任意现有导线、整个图形、项目中选定的图形或者整个项目。

第三节 原理图
Section 3 Schematic

编辑元件	edit component	链接元件	link component
插入/编辑辅元件	insert/edit child component	零件目录	part catalog
插入/编辑元件	insert/edit component	三相母线	3-phase bus wire
端子符号	terminal symbol	实时交互参考	real-time cross-referencing
多导线母线	multiple bus wire	替换元件	swap component
符号编译器	symbol builder	虚线	dashed line
回路编译器	circuit builder	选择开关	selector switch
回路配置	circuit configuration	移动回路	move circuit
回路元素	circuit element	元件标记	component tag
基础导线	underlying wire	元件描述	component description
继电器线圈	relay coil	元件注释	component annotation
快速移动元件	scoot a component	原理图符号	schematic symbol

1 You can go back to a component at any time and make changes. You can change description, tag, catalog, number, location code, terminal numbers, and rating values using the edit component tool.
你可以随时返回到元件并进行修改。可以使用"编辑元件"工具更改描述、标记、目录号、位置代号、端子号和额定值。

2 A circuit is made up of individual circuit elements and the wiring that connects them. Circuit Builder inserts a template drawing.
回路是由单独的回路元素以及连接这些回路元素的布线组成。回路编译器可以插入模板图形。

3 AutoCAD Electrical generates any of hundreds of different PLC I/O modules on demand, in various different graphical styles, all without a single, complete I/O module library symbol resident on the system. Modules adapt to the underlying ladder rung spacing, whatever that value might be. They can be stretched or broken into two or more pieces at insertion time.
AutoCAD Electrical 可以根据需要，采用多种不同的图形样式生成数百种不同的 PLC I/O 模块，这些模块在系统中都没有单一、完整的 I/O 模块库符号。模块会自动适应基本阶梯横档距，而不管取何间距值。插入模块时，还可以将模块拉伸或打断成两个或多个部分。

第四节 面板布局
Section 4 Panel Layout

半圆铭牌	half round nameplate	关联端子	associate terminals
备用端子	spare terminal	面板布局符号	panel layout symbols
插入附件	insert accessory	面板元件	panel component
插入示意图	insert footprints	名牌示意图	nameplate footprint
端子排	terminal strip	目录查找	catalog lookup
端子排编辑器	terminal strip editor	图标菜单	icon menu
复制端子特性	copy terminal block properties	位置代号	location code

第七章　电气设计 CAD 软件

下拉列表　drop-down
移动端子　move terminal

1 Using the AutoCAD Electrical panel layout tools, you can select from a list of schematic components. Place the footprint component directly into a panel layout. The footprint remains linked to the original schematic component, so you can perform bidirectional updating between schematic components and the associated footprint blocks.

通过 AutoCAD Electrical 面板布局工具,您可以从原理图元件的列表中进行选择。将示意图元件直接置于面板布局中。示意图仍与原始原理图元件保持链接,因此您可以在原理图元件和关联的示意图块之间执行双向更新。

2 Terminal strip editing is primarily used towards the end of the control system design cycle to expedite the labeling, numbering, and rearranging of terminals on a terminal strip.

端子排编辑主要在控制系统设计周期的结尾阶段使用,以提高在端子排上添加标签、编号和重新排列端子等操作的速度。

3 Terminal blocks connect devices that require quick disconnect or disassembly during product shipment, while at other times they can be used to distribute power to other devices. The terminal strip editor easily and quickly defines the locations for these connected devices during the system design process.

端子用来连接在产品装运期间需要快速分离或拆卸的装置,而有时还可以用来为其他装置配电。在系统设计过程中,可以使用端子排编辑器方便快捷地定义这些连接装置的位置。

第五节　生成报告
Section 5　Generating Reports

保存到文件　save to file
报告生成器　report generator
编辑 BOM 表报告　edit the BOM report
编辑报告　edit report
标准结算格式　normal tallied format
表格生成设置　table generation setup
电子表格　spreadsheet
更改报告格式　change report format
结算数量　tallied quantity
可选脚本文件　optional script file
目录号　catalog number
物料清单(BOM 表)　bill of material
制造商代号　manufacturer code
装配数量　subassembly quantity

1 Using AutoCAD Electrical, you can perform a project-wide extract of all BOM data found on your project drawing set. The data is extracted from the project database, matched with standard entries in the catalog database, and then additional fields are pulled from the catalog files. You can format this data into various report configurations and output to report files, export to a spreadsheet or database program, or place in an AutoCAD Electrical drawing.

使用 AutoCAD Electrical,您可以在整个项目范围内提取项目图形集上的所有 BOM 表数据。这些数据从项目数据库中提取出来,与目录数据库中的标准条目匹配,然后再从目录文件中提取附加字段。您可以将这些数据格式化为各种报告配置并输出到报告文件、电子表格或数据库程序中,或者放置到 AutoCAD Electrical 图形中。

2 Each AutoCAD Electrical report is customizable, from which data fields are reported and the order in which they appear to the justification of any column and the column labels.

每个 AutoCAD Electrical 报告都可以自定义,从中可以自定义报告的数据字段、字段的显示顺序以及列和列标签的对齐方式。

第八章　工控组态软件
Chapter 8　Software for Industrial Control System Configuration

ActiveX 控件脚本程序　ActiveX event script
安全　security
安装　installation
按键脚本程序　key script
版本　version
版权　copyright
报警　alarm
报警查询　alarm query
报警确认　alarm acknowledgment
报警限　alarm limit
变化率报警　rate-of-change alarm
变量值保存　value retention
标度变换原始值　scaling raw value
标识符字典　tagname dictionary
菜单　menu
操作系统　operating system
操作员系统标识符　operator system tag
测量单位　measurement unit
初值　initial value
窗口脚本程序　window script
垂直尺寸　vertical size
弹出窗口　popup window
当前报警　current alarm
导出　export
导入　import
登陆　log on
调试　debug
动画连接　animation link
动态分辨率选项　dynamic resolution option
动作脚本程序　action script
对话框　dialog box
多监视器系统　multi-monitor system
放大　enlarge
分布式报警系统　distributed alarm system
分布式应用　distributed application
分支语句　conditional branching
服务器　server
符号编辑器　symbol editor
复杂对象　complex object
覆盖窗口　overlay window

工厂操作员　plant operator
工具面板　tools panel
工位号　tag
画面　panel
集成开发环境　integrated development environment (IDE)
记录　log
技术支持　technical support
间接标识符　indirect tag
兼容性指导　compatible guide
检索　retrieve
脚本　script
脚本触发器　script trigger
脚本语言　script language
节点　node
开发系统　development system
可视化对象　visualization object
可视化元件　visualization element
客户端　client
控件　control
历史报警　historical alarm
历史趋势　historical trend
逻辑条件脚本程序　conditional script
密码用户　authenticating user
模板　template
模拟量报警　value alarm
逆时针旋转　rotate counterclockwise
配方管理器　recipe manager
偏差报警　deviation alarm
平板电脑　tablet PC
平方根标度变换　square root scaling
屏幕分辨率　screen resolution
嵌入式符号　embed symbol
趋势　trend
热备份　hot backup
热备份管理器　hot backup manager
人机接口　HMI/human machine interface
冗余　redundancy
实时趋势　real time trend
事件　event

第八章 工控组态软件

事件描述　event description
事件优先权　event priority
手动状态　manual mode
授权　licensing
数据变化脚本程序　data change script
数据采集与监控系统
SCADA　supervisory control and data acquisition
数据类型　type of data
数字量报警　discrete alarm
水平尺寸　horizontal size
顺时针旋转　rotate clockwise
死区　deadband
缩小　reduce
替代窗口　replacement window
条件分支　conditional branching
图形工具箱　graphic toolbox
图形接口　graphical interface
外部链接　external link
文档　documentation
文件夹路径　folder path
系统平台　system platform
线性标度变换　linear scaling
向导　wizard

项目　project
像素　pixel
选项　option
应用程序窗口　application window
应用系统标识符　application changed system tag
右对齐　right justified
访问级系统标识符　access level system tag
语言系统标识符　language system tag
远程引用　remote reference
远程桌面协议　remote desktop protocol
运行系统　runtime system
增补　supplementary component
智能符号　smart symbol
中央对齐　centered
终端服务　terminal service
主设备故障时启用的备份服务器　fail over server
属性　property
字符串　string
总貌　overview
组态　configuration
组态用户系统标识符　configure user system tag
左对齐　left justified

1 You can set an animation link for a text object to show the current value of a tag. This type of animation shows information to an operator like tank fill levels, machine on/off status, or alarm messages.
文本对象的动画链接可以显示工位标识符的当前值。这类动画链接将储罐液位、机器的开关状态、报警信息等显示给操作员。

2 Discrete tags are associated with process component properties whose values are represented by two possible Boolean states.
当过程装置的特性只有两种可能的逻辑状态时，可用离散工位标识符与之相关联。

3 The current tag value is compared to a target value. Then, the absolute value of the difference is compared to one or more alarm limits expressed as a percentage of the tag's possible value range.
将工位标识符的当前值与目标值进行比较,然后再将它们偏差的绝对值与一个或多个用量程百分比表达的报警限比较。

第九章 工业数据通信与网络技术
Chapter 9 Industrial Data Communication and Network Technology

第一节 概论
Section 1 Overview

0/1 序列　a sequence of zeros and ones
标准和协议　standard and protocol
测控系统　instrumentation and control system
工业以太网　industrial Ethernet
国际标准化组织
International Standard Organization (ISO)
局域网　local area network
开放系统互连参考模型
Open Systems Interconnection reference model
(OSI/RM)
数据链路层　data link layer
数据通信　data communication
数据通信设备　data communication equipment (DCE)
数据终端设备　data terminal equipment (DTE)
通信链路　communication link
网关　gateway (protocol converter)
物理层　physical layer
应用层　application layer

1 The challenge for the engineer and technician today is to make effective use of modern instrumentation and control systems and "smart" instruments. This is achieved by linking equipment such as PCs, programmable logic controllers (PLCs), and distributed control systems (DCSs) together with data communications systems.
对当代工程技术人员来说，挑战来自于如何有效地应用现代测控系统和"智能"仪表。测控系统是利用数据通信系统将各种测控设备相互连接，比如计算机、可编程序控制器和集散控制系统等。

2 Data communication is the transfer of information from one point to another.
数据通信指的是信息从一点向另一点传递。

3 Any communications system requires a transmitter to send information, a receiver to accept it and a link between the two.
任何通信系统都需要一个发射机来发送数据、一个接收机来接收数据，以及一个连接这两者的线路。

4 Traditionally, developers of software and hardware platforms have developed protocols, which only their products can use. In order to develop more integrated instrumentation and control systems, standardization of these communication protocols is required.
早期，软硬件平台的生产商开发了许多通信协议，但这些协议只能在其自己的产品上使用。为了便于和更多的测控系统相集成，需要对这些通信协议进行标准化。

5 The OSI model is useful in providing a universal framework for all communication systems. However, it does not define the actual protocol to be used at each layer.
OSI 模型为所有通信系统提供了一个有效和通用的框架。但它并未定义在每一层所使用的、具体的通信协议。

第二节 基本原理
Section 2 Basic Principles

差错检测　error detection
传输特性　transmission characteristics
单工、半双工和双工
simplex, half duplex and full duplex
数据编码　data coding
通用异步收发器
universal asynchronous receiver/transmitter (UART)

第九章 工业数据通信与网络技术

位、字节和字符 bits, bytes and characters
误码率 error rate
信噪比 signal to noise ratio (S/N ratio)
异步和同步系统 asynchronous and synchronous systems

1 A computer uses the binary numbering system, which has only two digits, 0 and 1.
计算机使用二进制计数系统,它只包括两个数:0 和 1。

2 The physical method of transferring data across a communication link varies according to the medium used.
传递数据的物理手段依照通信链路使用介质的不同而变化。

3 In any communications link connecting two devices, data can be sent in one of three communication modes. These are simplex, half duplex and full duplex.
无论任何通信链路,2 个设备之间的数据通信无外乎以下 3 种通信模式之一,即单工、半双工和双工。

4 An asynchronous system is one in which each character or byte is sent within a frame. The receiver does not start detection until it receives the first bit, known as the "start bit".
在异步通信过程中,每个字符或字节均在一帧数据中发送。接收端收到第 1 个数据位,也就是"起始位"后开始数据检测。

5 The basic principle of error detection is for the transmitter to compute a check character based on the original message content.
差错检测的基本原理是,发送端基于原始的信息内容计算出一个校验字符。

6 The start, stop and parity bits used in asynchronous transmission systems are usually physically generated by a standard integrated circuit (IC) chip that is part of the interface circuitry between the microprocessor bus and the line driver of the communications link.
异步传输系统中使用的起始位、停止位和校验位,通常采用标准的集成电路(IC)以物理方式产生,该电路是微处理器与通信链路的线路驱动器之间接口电路的一部分。

第三节 串行通信标准
Section 3 Serial Communication Standards

串行数据通信 serial data communication
电流环 current loop
电气信号特性 electrical signal characteristics
机械特性 mechanical characteristics
接口标准 interface standard
平衡与不平衡传输线路 balanced and unbalanced transmission lines
设备驱动 device drivers
通用串行总线 universal serial bus (USB)
拓扑 topology

1 An interface standard defines the electrical and mechanical details that allow equipment from different manufacturers to be connected and able to communicate.
接口标准对电气特性和机械特性给出了详细规定,从而使不同厂商的设备之间可以互连和通信。

2 Theoretically, unbalanced transmission should work well if the signal currents are small and the common conductor has very low impedance.
理论上讲,不平衡传输只在信号电流较小且公共导体阻抗很低的条件下正常工作。

3 The RS-485 interface standard is very useful for systems where several instruments or controllers may be connected on the same line.
如果系统要求在同一线路上连接多个仪表或控制器,RS-485 接口标准就显得非常有用。

4 Most of the systems encountered in data communications for instrumentation and control use some sort of software-based protocol in preference to hardware handshaking.
测控系统中常见的数据通信系统,更多地都使用基于软件的握手协议来替代硬件握手。

第四节 差错检测
Section 4　Error Detection

反馈差错控制　feedback error control
汉明码和汉明距离
Hamming codes and Hamming distance
块校验　block redundancy checks
前向差错控制　forward error correction
算术校验和　arithmetic checksum
信号衰减　signal attenuation
循环冗余校验　cyclic redundancy check (CRC)
延迟失真　delay distortion
字符校验　character redundancy checks

1 Signal attenuation is the decrease in signal amplitude, which occurs as a signal is propagated through a transmission medium.
信号衰减是指信号幅值的减小,这种现象发生在信号通过媒介的传输过程中。

2 Feedback error control is where the receiver is able to detect the presence of errors in the message sent by the transmitter. The detected error cannot be corrected but its presence is indicated.
采用反馈差错控制方法,接收端可以判断发送端发送的信息是否存在差错,但它只是确认差错的存在却不能加以纠正。

3 Before transmission of a character, the transmitter uses the agreed mechanism of even or odd parity to calculate the necessary parity bit to append to a character.
字符传输之前,发送端按照预先约定的奇偶校验机制计算出必要的校验位,并将其附加在原字符上。

4 A popular and very effective error checking mechanism is cyclic redundancy checking. The CRC is based upon a branch of mathematics called algebra theory, and is relatively simple to implement.
循环冗余校验(CRC)是一种常用且非常有效的差错检验方法。CRC 是基于数学的一个分支——代数理论,相对来说易于实现。

5 Forward error correction is where the receiver can not only detect the presence of errors in a message, but also reconstruct the message into what it believes to be the correct form.
采用前向纠错方法,接收端不仅能够判断信息是否存在差错,而且可以按照其认为正确的方式来重构该信息。

第五节 开放系统互连模型
Section 5　Open Systems Interface Model

表述层　presentation layer
传输层　transport layer
会话层　session layer
简化的 OSI 模型　simplified OSI model
数据包　packet
网络层　network layer
虚拟链路　virtual link

1 In digital data communications wiring together two or more devices is one of the first steps in establishing a network. As well as this hardware requirement, software must also be addressed.
在数字数据通信过程中,将 2 个或更多的硬件设备连接在一起是建立网络的第一步。除了满足硬件需求外,还需要有相应的软件支持。

2 It is important to realize that the OSI reference model is not a protocol or set of rules dictating how a protocol should be written but an overall framework in which to define protocols.
OSI 参考模型不是协议或规则集,它并不指明应该如何编写协议,而只是一个规范协议的总体框架,认识到这一点是很重要的。

3 For many industrial protocols the use of the full seven layers of the OSI model is inappropriate as the application may require a high-speed response. Hence a simplified OSI model is often preferred for industrial applications where time critical communications is more important than full communications functionality provided by the seven-layer model.
因为工业应用要求有快速的响应,所以对于许多工业通信协议来说,完整的 7 层 OSI 模型并不合适。

第九章　工业数据通信与网络技术

在工业应用中,对通信时间要求的重要性远高于7层模型所提供的完整的通信功能,所以才更多地采用一种简化的OSI模型。

第六节　工业数据通信协议
Section 6　Industrial Protocol

ASCII 协议　ASCII based protocol
Modbus 协议　Modbus protocol
读写命令　read and write command
功能码　function codes
请求帧　request frame
响应帧　response frame
信息格式　message format
远程终端　remote terminal unit (RTU)
主从模式　master/slave mode

1 The Modbus is accessed on the master/slave principle, the protocol providing for one master and up to 247 slaves. Only the master initiates a transaction.
Modbus 协议采用主/从原则,支持1个主站和最多247个从站,只有主站能发起通信请求。

2 The information in the message is the address of the intended receiver, what the receiver must do, the data needed to perform the action and a means of checking errors.
信息内容包括接收端的地址,接收端需要完成的任务,完成任务所需要的数据,以及差错检测的手段等。

3 All functions supported by the Modbus protocol are identified by an index number. They are designed as control commands for field instrumentation and actuators.
Modbus 协议支持的所有功能通过功能码(索引字)来加以区分。这些功能设计用作对于现场仪表和执行器的控制命令。

4 The most important protocols developed for internetworking are known as the TCP/IP Internet Protocols, usually abbreviated as TCP/IP.
为网络互联开发的最重要的协议是众所周知的TCP/IP互联网协议,通常缩写为TCP/IP。

第七节　局域网
Section 7　Local Area Network (LAN)

包交换　packet switching
城域网　metropolitan area network (MAN)
电路交换　circuit switching
广域网　wide area network (WAN)
介质访问控制　media access control
快速以太网　fast Ethernet
令牌传递　token passing
数据编码和传输　data encoding and transmission
网际互连　internetwork connection
网络操作系统　network operating system
星型、环型和总线型拓扑　star, ring and bus topologies
载波监听多路访问/碰撞检测　carrier sense multiple access/collision detection (CSMA/CD)
中继器、网桥、交换机和路由器　repeater, bridge, switch and router

1 A network is a system for interconnecting various devices, usually in such a way that all users have access to common resources (such as printers) and can communicate with each other.
网络是不同设备相互连接的一个系统,通常用来让所有用户访问共用的资源(比如打印机),并在相互之间进行通信。

2 In a circuit switched network, a connection is established between the two ends and maintained for the duration of the message exchange.
在电路交换网络中,通信两端建立连接并且在信息交换过程中保持连接。

3 A packet switched network does not establish a direct connection. Instead, the message is broken up into a series of packets or frames, sometimes known as protocol data units (PDU).
包交换网络并不建立直接的连接,而是将信息分拆为一系列的包或帧,有时也称为协议数据单元(PDU)。

4 Nodes in a ring network are connected node to node, ultimately forming a loop. The data flow is often arranged to be unidirectional, with each node passing data on to the next node and so on.

环型网络的节点与节点相连,最终形成一个闭环。数据流向大多为单向的,一个节点将数据传递到下一个节点,依此类推。

5 CSMA/CD is the simplest method of passing data, on a bus, between nodes that want to communicate in a peer-to-peer fashion.

对等网络的节点之间在总线上传递数据的最简单的方法就是CSMA/CD。

6 A node always waits until there is no carrier present on the line (indicating that no other node is transmitting) before sending a frame. It then monitors the cable as it transmits, and any difference between the signals it is sending and those it is receiving indicates that a collision has occurred (that is, another node has started transmission).

节点在发送数据帧之前一直在等待,直到确认在线路上没有载波信号存在(表明没有其他节点正在发送数据)。接下来,它在发送过程中还要监听电路上的信号,一旦在它发送和接收的信号之间出现任何差别,就说明碰撞产生了(也就是有其他节点开始发送数据)。

7 A special empty frame called a token is passed from one node to another, and a node can transmit data only when it holds the token. After confirming transmission of a data frame, the node generates a new token and sends it to the next node. This means that collisions cannot occur.

一个叫做令牌的空白帧从一个节点传到下一个节点,只有拥有令牌的节点才能发送数据。在确认数据帧发送结束后,这个节点产生一个新的令牌并将它交给下一个节点。这就意味着碰撞不可能发生了。

第十章 过程控制工程
Chapter 10 Process Control Engineering

第一节 绪论
Section 1 Introduction

闭环控制系统　closed-loop system
串联　in series
信号流图　information flow diagram
工艺流程图　process flow diagram
管道仪表流程图
　piping and instrument diagram/ P&ID
过程动态学　process dynamics
过程控制　process control

化学工程　chemical engineering
控制系统　control system
连续过程　continuous processes
上游过程　upstream process
设定值控制　set-point control
稳态　steady state
下游过程　downstream process

1 The measured value of the controlled variable is compared with a set value. The difference between these two values, the deviation, results in an output signal of the controller, acting on the correcting unit, which should eventually reduce or, preferably, eliminate the deviation.

受控变量的测量值与设定值比较。这两个值的偏差使控制器产生输出信号,该信号作用于校正单元,使偏差减小或消失。

2 The analysis of process and control loop dynamics is applied in the synthesis of control schemes.

过程和控制回路动态特性分析应用于控制方案综合。

3 Designing and operating an automated process so that it maintains specifications on profitability, quality, safety, environmental impact, etc., requires a close interaction between experts from different disciplines. These include, for example, computer-, process-, mechanical-, instrumentation- and control-engineers.

设计和操作自动化的生产过程,目的是保证效益、质量、安全和环保等指标,这需要包括计算机、工艺、机械、仪表和控制工程师等来自不同学科的专家之间的密切交流。

第二节 过程建模和过程检测控制仪表
Section 2 Modeling Process and Instruments

PID 控制算法　proportional-integral-derivative control algorithm
比例增益　proportional gain
变送器　transmitter
测量　measurement
传感器　sensor
磁滞　hysteresis
等百分比阀　equal percentage/increasing-sensitivity
调节阀　control valve
调节阀流量特性　valve characteristics
调节阀流通能力　control valve capacity
调节阀增益　valve gain
动量平衡　momentum balance
对数阀　exponential valve
阀杆　stem

阀门定位器　valve positioner
阀芯　valve plug/trim
非线性程度　nonlinearity
分辨率　resolution
工作流量特性　installed characteristics
管口　nozzle
管线　pipeline
过程模型　process model
横截面积　cross-sectional area
积分时间　integral time/reset time
精度　accuracy/error
开度　opening/shaft position
可调比　maximum rangeability
控制器　controller
快开阀　quick-opening/decreasing-sensitivity

理想流量特性　inherent characteristics
灵敏度　sensitivity
灵敏度漂移　sensitivity drift
零点漂移　zero drift
流量系数　flow coefficient
流速　flow rate
流体力学　fluid mechanics
满量程　full scale/full range
每分钟加仑　gpm
能量平衡　energy balance
偏差　error/deviation
气动阀（横隔膜）　diaphragm
气关阀　fail-open/air-to-close
气开阀　fail-close/air-to-open
手动状态　manual mode
输出流量　outlet flow
输入流量　inlet flow
数学模型　mathematical modeling

死区　deadband
微分时间　derivative time
无量纲的　dimensionless
物料平衡　material balance
线性度　linearity
线性阀　linear valve
校正动作　corrective action
斜率　slope
压力降　pressure drop
液位　level
仪表　instrument
阈值　threshold
振幅　amplitude
质量平衡　mass balance
重复性　repeatability/reproducibility
自动执行器　automatic actuator
自动状态　automatic mode

1 C_v is characteristic of a fully open valve. It's the horizontal flow in gpm of 60°F water under a 1 psid drop (there are no elevation contributions to the pressure drop used in this definition; it's all friction loss).

英制流量系数 C_v 定义：表示阀全开时的特性。阀水平放置，全开，阀两端压差为1磅/平方英寸（完全是摩擦引起的压降，无重力压降）时，通过温度为60°F 的水的流量。

2 Sensors are the "eyes" of control enabling one to "see" what is going on.

传感器是控制的眼睛，它使人能看见正在发生什么。

3 Alternatively, in those cases where particularly important measurements are not readily available then one can often infer these vital pieces of information from other observations. This leads to the idea of a "soft" or "virtual" sensor. We will see that this is one of the most powerful techniques in the control engineer's "bag of tools".

当重要的测量值不能测量时，一种可选办法是从其他可测的信息来推导出这些重要的信息。这引出了"软"或"虚"传感器的概念。我们将看到这将是控制工程师的工具包中最强有力的工具之一。

第三节　单回路控制系统的工程设计
Section 3　Single Loop Control System and its Design

被测变量　measured variable
比例度　proportional band
变系数　variable-coefficient
常系数　constant-coefficient
超调量　overshoot
初始状态　initial conditions
传递函数　transfer functions
传感器　transducer/sensor
纯滞后时间　dead time
单回路反馈控制　single-loop feedback control
单位阶跃　unit step
调节器整定　controller tuning

动态特性　dynamic behavior
二次方程　quadratic equation
二阶的　second-order
发散振荡　increasing oscillation
非线性函数的近似　approximating a nonlinear function
非最小相位特性　nonminimum-phase behavior
峰值　peak
复平面　complex plane
复数　complex number
傅立叶变换　Fourier transforms
干扰　disturbance

过程静态增益　static process gain	泰勒级数展开　representing functions by Taylor series
过阻尼系统　over damped systems	特征方程　characteristic equation
极点　pole	微分方程　differential equations
矩形脉冲　rectangular pulse	稳定性　stability
控制变量　manipulated variable	线性化　linearization
控制方案　control scheme	线性模型　linear model
控制性能指标　measures of control performance	线性系统稳定判据　criterion for stability of linear systems
控制质量　control quality	响应时间　response time
拉普拉斯变换　Laplace transforms	斜坡(信号)　ramp
临界阻尼的　critically damped	一阶惯性环节　first-order lag
零点　zero	一阶过程　first-order process
脉冲干扰　impulse disturbance	因变量　dependent variable
偏差变量　deviation variables	暂态响应　transient response
频率响应　frequency response	振荡特性　oscillatory behavior
欠阻尼的　undercritically damped	振幅　amplitude of oscillation
上升时间　rise time	正弦输入　sine input
设定值　setpoint variable	执行元件　final control element
时变系统　time-varying system	指数　exponential
时间常数　time constant	终值　ultimate value
时间延迟　time delay	周期　period
时域　time domain	自变量　independent variable
受控变量　controlled variable/process value	自然频率　natural period
衰减比　decay ratio	

1 Feedback control means measuring the controlled variable (an output), comparing that measurement to the set point (desired value), and acting in response the error (difference between set point and controlled variable) by adjusting the manipulated variable (an input).

反馈控制是测量受控变量(输出),将它与设定值(期望值)比较,通过改变控制变量(输出)来对偏差(设定值与受控变量间的差值)做出响应。

2 Control is concerned with finding technically, environmentally and commercially feasible ways of acting on a technological system to control its outputs to desired values whilst ensuring a desired level of performance.

控制就是要找到一种在技术上、环保上和商业上都可行的方法,作用于受控系统,使系统的输出被控制在设定值上,同时确保期望的性能指标。

3 Fundamental to control engineering is the concept of inversion. Inversion can be achieved by a feedback architecture.

控制工程的基本原理是转换。转换可以通过反馈方案实现。

4 The three elements of PID control each have a distinct personality, but they can be made to work together. Loop tuning becomes easier when you understand the interactions.

PID控制的比例、积分和微分作用都有各自的特点,但是它们可以组合在一起工作。如果你理解它们之间的相互影响,回路整定就变得比较简单。

第四节　复杂过程控制系统
Section 4　Complex Control System

安全装置　foolproof	波动　fluctuation
保护装置　protection device	玻璃液面计　gage glass
比值控制　ratio control	残液　raffinate

操作成本　operation cost
测量信号　measurement signal
超驰控制　override control
超前滞后环节　lead/lag elements
串级控制　cascade control
催化剂　catalyst
催化裂化装置　catalytic cracking unit
萃取塔　extraction column
低选器　low value selector
电/气转换器　current/pressure converter
电磁阀　solenoid valve
电磁流量计　magnetic flowmeter
多级泵　multiple impeller pump
法兰　flange
放热的　exothermic
废催化剂　fouled catalyst
分布式控制系统　distributed control system(DCS)
分程控制　split range control
分馏塔　fractional distillation
副调节器　slave controller/secondary controller
副回路　slave control loop/secondary control loop
高选器　high value selector
固定床操作　fixed-bed operation
固定流化床　fluid fixed bed
管式加热炉　tube still
管线调和　in-line blending
罐的呼吸　exhalation of tank
过程设备　process apparatus
过热器　steam superheater
过热蒸汽　gaseous steam
呼吸阀　exhalation valve
回流　reflux
火炬　flare stack
计量泵　metering pump
加氢精制　hydrofining
加氢裂化　hydrocracking
加热炉　furnace
加热盘管　heating coil
加热器　heat booster
甲烷化　methanize
减压塔　vacuum column
结垢　fouling
截止阀　globe valve
解耦控制　decoupling control
精馏塔　distillation column
可靠性　reliability

空气-燃料混合物　fuel mixture
孔板　orifice
裂解炉　cracker
临界速度　critical speed
流化催化裂化　fluid catalytic cracking
流体　fluid
馏分　fraction
漏气　fizz
炉膛　firebox
排气　exhaust/exit gas
气鼓　drum
气提塔　stripper
前馈控制　feedforward control
前馈作用　anticipating action
强制通风　forced draft
全回流　infinite reflux
燃料供给　fuel supply
燃料喷嘴　fuel injector
燃料输送泵　fuel transfer pump
燃料油　fuel oil
热电偶　thermocouple
热交换器　heat exchanger
容器　vessel
溶剂　solvent
三通阀　T-valve
闪蒸罐　flash drum
省煤器　economizer
施工图　execution drawing
实际装置　full-scale plant
试验装置　pilot-scale plant
输出流量　outgoing flow
顺流　forward current
塔板　tray
涡流　erratic flow
吸入流量　incoming flow
现场控制总线　field control bus
相对增益　relative gain
压缩机　compressor
烟道　flue
烟气　flue gas
扬程　lift of pump
一阶反应　first order reaction
优化控制　optimizing control
原料　raw material
圆顶罐　globe-roof tank
约束值　constraint value

第十章 过程控制工程

闸阀　gate valve
蒸发器　evaporator
蒸汽锅炉　steam boiler
重整装置　reformer
主调节器　master controller/primary controller
主回路　master control loop
转速计　tachometer
自动调节阀/带伺服电机　pilot valve/motor valve
自动加料罐　gravity tank

1 Cascade control is still feedback control, performed with conventional PID control algorithms. The control scheme features a new secondary loop within the original primary loop.
串级控制仍然是用 PID 控制算法实现的反馈控制，它的特点是在原来的主回路内又增加了一个副回路。

2 It frequently happens that what we have designated as disturbances in a given control loop, are signals originating in other loops, and vice versa. This phenomenon is known as interaction or coupling. In some cases, interaction can be ignored, either because the coupling signals are weak or because a clear time scale or, frequency scale separation exists. However, in other cases it may be necessary to consider all signals simultaneously. This leads us to consider multi-input multi-output (or MIMO) architectures.
在控制系统中一直被认为是干扰的信号往往来自于其他回路。这种现象被认为是相互作用或耦合。有些情况，耦合可以忽略，忽略的原因是耦合信号太弱，或者是在时域或频域上可以分离。然而，有时必须同时考虑所有信号，这导致我们要考虑多输入多输出（或 MIMO）方案。

3 The jacket temperature of a chemical reactor can be controlled by cooling water valve and steam. The controller actuates two control valves in split range: only after the water valve has been closed is the steam valve opened, and vice versa.
冷却水和蒸汽可以控制化学反应器的夹套温度，控制器以分程的方式控制两个调节阀：只有冷却水阀关闭后，蒸汽阀才能打开，反之亦然。

第十一章 集散控制系统与现场总线技术
Chapter 11 Distributed Control System and Fieldbus Technology

第一节 概述
Section 1 Overview

测控系统　instrumentation and control system
工控机　industrial PC (IPC)
集散控制系统　distributed control system (DCS)
集中管理　central management
人机界面　human machine interface (HMI)

可编程序控制器　programmable logic controller (PLC)
实时分布式控制　real-time distributed control
智能仪表　smart instrument
组态软件　configuration software

1 A distributed control system (DCS) refers to a control system usually of a manufacturing system, process or any kind of dynamic system, in which the controller elements are not central in location but are distributed throughout the system with each component sub-system controlled by one or more controllers.

集散控制系统是一种应用于制造业、生产过程或任何一种动态系统的控制系统。集散控制系统的控制器在位置上是分散而不是集中的，每个控制单元子系统由一个或多个控制器控制。

2 A DCS typically uses custom designed processors as controllers and uses both proprietary interconnections and communications protocol for communication.

集散控制系统通常采用专门设计的处理器作为控制器，并且采用其专属的互连方式和通信协议进行通信。

3 Distributed control systems (DCSs) are dedicated systems used to control manufacturing processes that are continuous or batch-oriented, such as oil refining, petrochemicals, central station power generation, fertilizers, pharmaceuticals, food and beverage manufacturing, cement production, steelmaking, and papermaking.

集散控制系统是专门用来控制各类连续或批量生产过程的系统，例如炼油、石化、发电、化肥、制药、食品、饮料、水泥、炼钢和造纸等。

4 The DCS largely came about due to the increased availability of microcomputers and the proliferation of microprocessors in the world of process control.

集散控制系统在过程控制领域的大量应用，是得益于微计算机更加易用以及微处理器的普及。

5 Fieldbus is an industrial network system for real-time distributed control. It is a way to connect instruments in a manufacturing plant.

现场总线是一种应用于实时分布式控制的工业网络系统，是在工厂连接现场仪表的一种方式。

第二节 DCS 的结构、硬件和通信
Section 2 Architecture, Hardware and Communication

操作员站　operator station
高速数据通道　data highway
工程师站　engineer station
监督式控制　supervisory control
模拟信号　analog signals
冗余的处理器　redundant processors

输入输出模块　I/O modules
数据编码　data coding
数据采集单元　data acquisition unit
数字信号　digital signals
现场控制单元　local control unit
直接数字控制　direct digital control (DDC)

1 Measurements are transmitted to computer and control signals are sent from computer to control valves at specific time interval known as sampling time.

第十一章 集散控制系统与现场总线技术

检测信号送到计算机,控制信号按一定的时间间隔(称作采样周期),从计算机送出以控制阀门开度。

2 Direct digital control, by contrast, requires that all control actions be carried out by the digital computer.

相反,直接数字控制中所有控制动作均由数字计算机执行。

3 The modern DCS are equipped with optimization, high-performance model-building and control software as options. Therefore, an imaginative engineer who has theoretical background on modern control systems can quickly configure the DCS network to implement high performance controllers.

现代的集散控制系统可以选配优化、高性能建模和控制软件。因此,颇具想象力的工程师如果具有现代控制系统的理论背景,可以很快地组态 DCS 网络来实现高性能的控制器。

4 Many field sensors naturally produce analog voltage or current signals. For this reason transducers that convert analog signals to digital signals (A/D) and vice verse (D/A) are used as interface between the analog and digital elements of the modern control system.

许多现场仪表自身输出模拟电压或电流信号,因此,在现代控制系统中,使用转换器件作为模拟信号和数字信号间的接口,将模拟信号转换为数字信号(A/D),或者反过来将数字信号转换为模拟信号(D/A)。

5 For each point, a separate alarm priority can be specified for each alarm (for example, PV high alarm can be low priority but PV high high alarm can be emergency).

对于每个数据点,可以分别为每个报警设定报警优先级(比如,当前值 PV 的高报警可以是低优先级,而高高报警可以设为紧急级别)。

6 The input/output devices (I/O) can be integral with the controller or located remotely via a field network.

输入输出模块既可以和控制器集成在一起,也可以位于远端并通过现场网络与控制器相连接。

第三节　软件设计
Section 3　Software Design

常规控制点	regulatory control point	面向控制的编程语言	control-oriented programming languages
程序模板	template routines	神经网络和模糊控制	neural networks and fuzzy control
定值控制	setpoint control	线性变换	linear conversion
功能块	function blocks		
控制算法	control algorithms		
模式切换	mode switching		

1 The operator interface is generally a terminal upon which the operator can communicate with the system. Such terminals usually permit displaying graphical information.

操作员界面是操作员和系统通信的终端,通常用来显示图形化信息。

2 To make the best use of a DCS system, an advance control strategy or supervisory optimization can be incorporated in the main host computer.

为了充分利用集散控制系统,可以在主计算机上运行先进控制策略或监督优化软件。

3 In many cases, the user is able to utilize the template routines supplied by the vendor, and is required only to duplicate these routines and interconnect them to fit his own application purposes.

大多数情况下,用户可以直接使用厂家提供的程序模板,只需拷贝这些模版然后根据自己具体的应用功能进行内部连接即可。

4 Similarly, many control algorithm developers design a special interface to allow incorporating their own control programs into most of the commercial DCS network.

同时,许多控制算法开发商设计出了特殊接口,用户可以通过它将自主开发的控制程序嵌入到大多数商品化的集散控制系统当中。

第四节　现场总线技术
Section 4　Fieldbus Technology

电子控制单元　electronic control unit (ECU)
多主广播模式串行总线
multi-master broadcast serial bus
基于信息的协议　message-based protocol
结束分隔符　end delimiter
菊花链　daisy-chain
开放结构控制器　open architecture controller

扩展帧格式　extended frame format
曼彻斯特编码电流调制
manchester-encoded current modulation
双绞线电缆　twisted pair cables
与安全相关　safety-relevant
总线仲裁　bus arbitration

1 This would be the equivalent of the currently used 4-20 mA communication scheme which requires that each device has its own communication point at the controller level, while the fieldbus is the equivalent of the current LAN-type connections, which require only one communication point at the controller level and allow multiple (hundreds) of analog and digital points to be connected at the same time.

这种方案等同于目前常用的 4-20mA 通信，它要求每个设备在控制级都具有自己的通信节点；而现场总线方案等同于当前的局域网连接类型，它在控制级只有一个通信节点，可以同时连接很多个（数百）模拟和数字信号点。

2 Although fieldbus technology has been around since 1988, with the completion of the ISA S50. 02 standard, the development of the international standard took many years.

尽管现场总线技术早在 1988 年就已经出现，其标志是 ISA S50. 02 标准的完成，但国际标准的发展却花费了很多年。

3 The physical medium and the physical layer standards fully describe, in detail, the implementation of bit timing, synchronization, encoding/decoding, band rate, bus length and the physical connection of the transceiver to the communication wires.

物理介质和物理层标准全面详细地描述了如何实现位定时、同步、编码/解码、波特率、总线长度以及收发器与通信线路的物理连接形式等。

4 The security layer FDL (Field bus Data Link) works with a hybrid access method that combines token passing with a master-slave method. In a PROFIBUS DP network, the controllers or process control systems are the masters and the sensors and actuators are the slaves.

安全层的 FDL（现场总线数据链接）协议采用混合访问机制，包含令牌传递和主从两种类型。在 PROFIBUS DP 网络中，控制器或过程控制系统作为主站，传感器和执行器作为从站。

5 Since Modbus is a master/slave protocol, there is no way for a field device to "report by exception" (except over Ethernet TCP/IP) - the master node must routinely poll each field device, and look for changes in the data.

因为 Modbus 协议采用主从机制，所以现场设备不能主动报告意外状态（Ethernet TCP/IP 除外），主站节点必须按部就班第轮询每个现场设备，以查看数据的变化情况。

6 CAN bus (for controller area network) is a vehicle bus standard designed to allow microcontrollers and devices to communicate with each other within a vehicle without a host computer. CAN bus is a message-based protocol, designed specifically for automotive applications but now also used in other areas such as industrial automation and medical equipment.

CAN（控制器局域网的简写）总线是一种汽车总线标准，设计用于在一辆汽车内、没有主计算机的条件下实现微处理器和设备之间的相互通信。

7 Industrial Ethernet (IE) refers to the use of the Ethernet family of computer network technologies in an industrial environment, for automation and process control. A number of techniques are used to adapt Ethernet for the needs of industrial processes, which require real time behavior.

工业以太网(IE)将计算机网络中的以太网技术应用于自动化和过程控制等工业环境。它采用了许多技术以使以太网满足工业过程所要求的实时性。

8 One or more fieldbus standards may predominate in future and others may become obsolete. This increases the investment risk when implementing fieldbus.

一种或多种现场总线标准未来可能会成为主宰,其他的可能会过时,这一点增加了采用现场总线的投资风险。

第十二章 计算机控制技术
Chapter 12　Computer Control Techniques

第一节　绪论
Section 1　Introduction

被测参数　measured parameter	连续时间信号　continuous time signal
标准电信号　standard electric signal	模糊控制系统　fuzzy control system
标准总线　standard bus	模拟控制系统　analog control system
操作站　operator station	模拟信号　analog signal
操作指导　operator guide	平均无故障时间　mean time between failures (MTBF)
电动执行器　electric actuator	气动执行器　pneumatic actuator
多级计算机控制　hierarchical computer control	生产过程　production process
工程师站　engineering station	生产计划与调度　production planning and scheduling
工业过程控制　industrial process control	实时时钟　real-time clock
工业生产对象　industrial production object	外部通用设备　exterior universal device
工业自动化系统　industrial automation system	系统辨识　system identification
管理信息系统　management information system	现场总线控制系统　field bus control system
集散控制系统　distributed control systems (DCS)	学习控制系统　learning control system
计算机集成制造系统　computer integrated manufacturing system	直接数字控制　direct digital control (DDC)
计算机监督控制　supervisory computer control (SCC)	智能控制系统　intelligent control system
计算机控制技术　computer control technique	智能仪表　intelligent instrument/smart instrument
计算机控制系统　computer controlled system	中断设备　interrupt device
监控管理程序　supervisory control management program	中央控制单元　central control unit
检测元件　detecting element	专家控制系统　expert control system
聚合装置　polymerization unit	自动整定　automatic tuning
开放式控制系统　open control system	自适应控制　adaptive control
可编程逻辑控制器　programmable logic controller	总线工业控制机　bus industrial control computer
离散时间信号　discrete time signal	最优控制　optimal control

1 DCS systems generally consist of process stations, which controlling the process; operator stations, where process operators monitor activities; and various auxiliary stations.
分布式控制系统通常包括控制过程的过程站，具有操作监视作用的操作站和各种辅助的站点。

2 Based on the dramatic developments in the past, it is tempting to speculate about the future. There are four areas that are important for the development of computer process control:
- process knowledge
- measurement technology
- computer technology
- control theory

在已有的引人注目的发展基础上，展望未来，前景诱人。要发展计算机过程控制，下面四个领域非常重要：
- 过程知识
- 测量技术
- 计算机技术
- 控制理论

第二节 模拟量输入输出通道接口技术
Section 2 Interface Technology for Analog Input/Output Channel

采样保持电路	sampling-and-hold circuit	模拟量输出通道	analog output channel
采样定理	sampling theorem	模拟量输入通道	analog input channel
采样频率	sampling frequency	软件延时方式	software delay mode
采样系统	sampling system	时不变系统	time-invariant systems
采样周期	sampling period/sampling interval	数字地	digital earth
参考电平	reference voltage	数字量输出通道	digital output channel
查询方式	query mode	数字量输入通道	digital input channel
多速率采样	multirate sampling	随机采样	random sampling
解码网络	decoding network	通道与接口	channel and interface
离散时间系统	discrete-time system	直接存储器存取	direct memory access
连续时间系统	continuous-time system	中断方式	interrupting mode
模拟地	analog earth	周期采样	periodic sampling

1 In the context of control and communication, sampling means that a continuous-time signal is replaced by a sequence of numbers, which represents the values of the signal at certain times.

在控制和通信领域,采样则意味着一连续时间信号由一个数值序列所代替,这个数列代表了某些时刻的信号值。

2 Shannons sampling theorem: A continuous-time signal with a Fourier transform that is zero outside the interval $(-\omega_0, \omega_0)$ is given uniquely by its values in equidistant points if the sampling frequency is higher than $2\omega_0$. The continuous-time signal can be computed from the sampled signal by the interpolation formula

$$f(t) = \sum_{k=-\infty}^{\infty} f(kh) \frac{\sin[\omega_s(t-kh)/2]}{\omega_s(t-kh)/2}$$
$$= \sum_{k=-\infty}^{\infty} f(kh) \operatorname{sinc} \frac{\omega_s(t-kh)}{2}$$

where ω_s is the sampling angular frequency in radians per second (rad/s).

香农采样定理:如果一个连续时间信号的傅立叶变换在$(-\omega_0,\omega_0)$之外等于零,那么当采样频率大于$2\omega_0$时,该信号可由它的等时间距离点上的值所唯一确定。这时应用插值公式:

$$f(t) = \sum_{k=-\infty}^{\infty} f(kh) \frac{\sin[\omega_s(t-kh)/2]}{\omega_s(t-kh)/2}$$
$$= \sum_{k=-\infty}^{\infty} f(kh) \operatorname{sinc} \frac{\omega_s(t-kh)}{2}$$

就能由采样信号算出这个连续时间信号。式中ω_s是采样角频率,单位为弧度/秒(rad/s)。

3 Mass markets such as automotive electronics have also led to the development of special-purpose computers, called micro controllers, in which a standard computer chip has been augmented with A-D and D-A converters, registers, and other features that make it easy to interface with physical equipment.

巨大的市场潜力,如汽车电子自动化的需求使得一些称为微控制器的特殊用途计算机得到了发展,在这些微控制器中,一块标准的计算机芯片已增添了模数转换器和数模转换器、寄存器和其他特性,这样使微控制器可以方便地与物理设备接口。

4 It is also possible to use different sampling periods for different loops in a system. This is called multirate sampling.

在一个系统的不同回路中,采用不同的采样周期,这种采样称为多速率采样。

第三节 人机接口技术
Section 3　Man-machine Interface Technology

点阵式显示器　matrix display
动态显示　dynamic display
抖动　dithering
发光二极管　light emitting diode (LED)
功能键　function key
光柱显示器　light columniation display
记录仪表　recording instrument
静态显示　static display
矩阵键盘　matrix keyboard
连击　double hit
描述法　describing method
数码管　digital displaying diode
显示仪表　displaying instrument
液晶显示器　liquid crystal display (LCD)
硬件译码　hardware coding
重击　repeating hit

1 Man-machine interface is the region where a human being comes in contact with a machine, or the relationship between a human and a machine. One must ensure that the interface is designed properly for the capabilities and limitations of the users.

人机接口是指人和机器接触的区域或是人和机器之间的联系。必须保证接口的设计要恰当地满足用户的接受能力和局限性。

2 It becomes clear that process control puts special demands on computers. The need to respond quickly to demands from the process leads to the development of the interruption feature, which is a special hardware device that allows an external event to interrupt the computer in its current work so that it can respond to more urgent process tasks.

很明显，过程控制向计算机提出了许多特殊的要求，需要它对各种过程命令做出迅速响应，从而导致发展中断设备，这是一种特殊的硬件装置，它允许外部事件中断计算机的当前工作，以便计算机能够对更紧迫的过程任务做出响应。

第四节 通用的控制程序设计
Section 4　Universal Control Program Design

报警程序设计　alarm program design
闭环脉冲宽度调速系统　closed-loop pulse width speed regulation system
变速控制　shift gear control
测速发电机　speed indicating generator
大功率场效应管　high-power field effect transistor
单向可控硅　unilateral controlled silicon
电磁阀　electro-magnetic valve
电子开关　electronic switch
定时器　timer
反向转动　negative directional turn
方向控制　directional control
固态继电器　solid state relay
光电隔离技术　photoelectric isolation
计数法　counting method
交流电动机　alternating motor
交流继电器　alternating current relay
开关量　switch variable
可控硅接口技术　controlled silicon interface technology
脉冲宽度调制　pulse width modulation (PWM)
平均速度　average speed
驱动器　driver
软件延时法　software delay method
数字式转速传感器　digital rotate speed sensor
双向可控硅　triac
小功率直流电动机　low-power direct current motor
越限报警　beyond range alarm
占空比　duty cycle/duty ratio
正向转动　forward directional turn
直流继电器　direct current relay
专用接口板　special interface board
专用接口芯片　special interface chip

1 Pulse width modulation, abbreviated as PWM, is a method of transmitting information on a series of pulses. The data being transmitted are encoded on the width of these pulses to control the amount of power being sent to a load.

脉冲宽度调制(简写PWM),是以串行脉冲的方式传递信息的一种方法。传递的数据编码为脉冲的宽度来控制传递给负载的能量。

2 Pulse width modulation is a modulation technique for generating variable width pulses to represent the amplitude of an input analog signal or wave. The popular applications of pulse width modulation are in power delivery, voltage regulation and amplification and audio effects.

脉冲宽度调制是一种调制技术,产生可变宽度的脉冲,描述输入模拟信号或波形的幅度。脉冲宽度调制广泛应用在电力传输、电压调制、放大和音频效果方面。

3 Industrial automation systems traditionally have two components, controllers and relay logic. Relays are used to sequence operations such as startup and shutdown. They are also used to ensure safety of the operations by providing interlocks.

工业自动化系统在传统上有两个部分:控制器和继电器逻辑电路。继电电路通常用于顺序操作,如开和关,同时它通过提供连锁来保护操作的安全。

第五节　总线接口技术
Section 5　Bus Interface Technology

半双工方式　half duplex mode
并行总线　parallel bus
差错控制技术　error control technology
串行通信　serial communication
串行通信标准总线　serial communication standard bus
串行总线　serial bus
单工方式　simplex mode
等待请求　waiting request
地址锁存允许　address locking enable
地址总线　address bus
电平转换芯片　electric level conversion chip
电气特性　electric performance
定时控制　timing control
发送数据　transmitting data
幅移键控法　amplitude shift key
工艺流程　production process
国际标准化组织　International Standard Organization (ISO)
国际电工委员会　International Electrical Committee (IEC)
过程现场总线　process field bus
混合纠错　hybrid error correction
机械特性　mechanical performance
纠错编码　error correction coding
局部操作网络　local operating network
开放系统互连参考模型　open system interconnection reference model
可寻址远程传感器数据通路　highway addressable remote transducer
控制器局域网络　controller area network

控制总线　control bus
路由器　router
模块式结构　modularization structure
频移键控法　frequency shift key
奇偶校验　parity check
前向纠错方式　forward error correction
清除发送　clear to send
请求发送信号　request to send
全双工方式　full duplex mode
神经元　neuron
神经元专用芯片　neuron special chip
数据传送方式　data transfer mode
数据通信设备　data communication equipment
数据载波检测信号　data carrier detect
数据终端就绪信号　data terminal ready
数据准备就绪　data set ready
数据总线　data bus
同步通信　synchronous communication
网络　network
现场总线基金会　field foundation (FF)
相移键控法　phase shift key
协议　protocol
信号调制　signal modulation
选择接口控制　selection interface control
循环冗余校验　cyclic redundancy check
仪表装置　instrument equipment
异步通信　asynchronous communication
载波侦听多路访问/介质访问控制　carrier sense multiple access
振铃指示信号　ring indication
智能变送器　intelligent transducer

智能现场设备　intelligent field device　　　　自动重发控制　automatic repeat request
中断请求　interrupt request　　　　　　　　　总线请求　bus request
中断响应　interrupt response　　　　　　　　 总线响应　bus response
自动化技术　automation technology

1 Fieldbus is a generic-term which describes a new digital communications network which will be used in industry to replace the existing 4 - 20mA analog signal.
现场总线是一个描述新型数字通信网络的常用术语,可以用在工业上取代4-20mA模拟信号。

2 Fieldbus is a digital, bi-directional, multidrop, serial-bus communications network used to link isolated field devices, such as controllers, transducers, actuators and sensors. Each field device has low cost computing power installed in it, making each device a "smart" device.
现场总线一个数字化的、双向的、多节点的、串行总线的通信网络,用于连接分散的现场设备,如控制器、变送器、执行器和传感器。每个现场设备内部有低成本的运算单元,使其成为智能设备。

3 Each device will be able to execute simple functions on its own such as diagnostic, control, and maintenance functions as well as providing bi-directional communication capabilities. With these devices not only will the engineers be able to access the field devices, but they are also able to communicate with other field devices.
每个设备都能执行简单的功能,如诊断、控制、维护以及双向的通信能力。使用这些设备,工程师不仅能访问现场设备,而且能与其他的现场设备通信。

4 In essence, fieldbus will replace centralized control networks with distributed-control networks. Therefore, fieldbus is much more than a replacement for the 4 - 20mA analog signal standard.
本质上现场总线用分散化的控制网络取代集中化的控制网络。因此,现场总线不仅仅是替代4-20mA模拟信号标准。

第六节　过程控制的数据处理
Section 6　Data Processing Methods for the Process Control

标度转换　scale conversion　　　　　　　　　　量程自动转换　span automatic conversion
查表技术　checking list technology　　　　　　 模拟滤波　analog filtering
程序判断滤波　program judge filtering　　　　　容错技术　fault tolerant technology
低通数字滤波　low pass digital filtering　　　　数据处理　data processing
对分查表法　bisection look up method　　　　　数字滤波　digital filtering
非线性参数　nonlinear parameter　　　　　　　 顺序查表法　sequential look up method
复合数字滤波　compound digital filtering　　　　算术平均滤波　arithmetic average filtering
工业控制组态软件　　　　　　　　　　　　　　限幅滤波　clipped filtering
industrial control configuration software　　　　限速滤波　velocity limiting filtering
滑动平均滤波　moving average filtering　　　　 智能化仪器　intellectualized instrument
计算查表法　computing look up table method　　中值滤波　median filtering
加权平均滤波　weighted average filtering　　　　自动校正　self tuning
可编程增益放大器　programmable gain amplifier

1 A low-pass filter blocks high frequencies and passes low frequencies. It would be most desirable if a low-pass filter had a characteristic such that all signals with frequency above some critical value are simply rejected. A high-pass filter passes high frequencies (no rejection) and blocks (rejects) low frequencies.
低通滤波器使低频信号通过、高频信号阻断。如果低通滤波器能够使超过某个转折频率的所有频率信号截止,那将更完美。高通滤波器使高频信号通过,阻止低频信号通过。

2 Using computers to implement controller has great advantages. Many of the difficulties with analog implementation can be avoided. For example, there are no problems with accuracy or drift of the components. It is very easy to have sophisticated calculations in the control law, including logic and nonlinear functions.

用计算机做控制器有很多的优点,可以克服一些模拟控制器的困难,如没有精度或者元件的漂移问题。也很容易在控制规律中完成精密的计算,包括逻辑的和非线性的函数。

3 The ideal filters pass all signals in the passband with a gain of unity and completely reject all signals in the stopband. The cutoff frequencies are typically defined by the points where the magnitude drops to $1/\sqrt{2}$. In real filters, the magnitude can not change discontinuously, so there is some transition band between the passband and the stopband.

理想的滤波器在通频带中以同一增益通过所有信号,在衰减带中禁止所有信号通过。剪切频率定义在幅值下降为 $1/\sqrt{2}$ 处。实际的滤波器中,幅值不能间断地变化,所以在通频带和衰减带之间有一个过渡带。

第七节 数字 PID 算法
Section 7 Digital PID Algorithm

s 平面　s-plane
z 变换　z transform
z 平面　z-plane
临界比例度法　ultimative-sensitivity method
传递函数　transfer function
位置式　position form
参数整定　parameter tuning
反作用　negative action
响应曲线法　transient-response method
增量式　incremental form/velocity form
微分时间　derivative time
手动控制　manual control
抗积分饱和　antireset windup
拉普拉斯变换　Laplace transform
拉普拉斯逆变换　inverse Laplace transform
控制回路　controlled loop
控制方式　control mode
控制阀　control valve
数字控制器　digital controller
数字控制算法　digital control algorithm
无扰动切换　bumpless transfer
条件积分　conditional integration

欠阻尼　underdamped
正作用　positive action
比例增益　proportional gain
流程图　flowchart
流量　flow rate
测量值　measured value
液位　liquid level
离散控制系统　discrete-time control system
积分时间　integration time / reset time
积分补偿　integration offset
积分饱和　reset windup/integrator saturation
程序和顺序控制　programmable and sequential control
给定值　reference value/set point
自动控制　automatic control
被控参数　controlled parameter
被控对象　controlled object
调节过程　control process
过阻尼　overdamped
连续控制系统　continuous-time control system
速度控制　velocity control
静态误差　steady-state error

1 A pure derivative cannot, and should not be, implemented, because it will give a very large amplification of measurement noise. The gain of the derivative must thus be limited. This can be done by approximating the transfer function as follows:

$$sT_d \approx \frac{sT_d}{1+sT_d/N}$$

The transfer function on the right approximates the derivative well at low frequencies but the gain is limited to N at high frequencies. N is typically in the range of 3 to 20.

一个纯微分不能实现,或者最好不要实现,因为这将导致噪声被显著放大。因此微分增益必须受到限制。为此可把传递函数 sT_d 近似为:

$$sT_d \approx \frac{sT_d}{1+sT_d/N}$$

上式右边的传递函数在低频段内能很好地近似微分,而在高频段就把增益限制到 N,N 的范围通常在 3 到 20 之内。

2 Many practical control problems are solved by PID-controller. The "textbook" version of the PID-controller can be described by the following equation

$$u(t) = K\left(e(t) + \frac{1}{T_i}\int_0^t e(s)ds + T_d\frac{de(t)}{dt}\right)$$

where error e is the difference between command signals u_c (the set point) and process output y (the measured variable). K is the *gain* or *proportional gain* of the controller, T_i the *integration time* or *reset time*, and T_d the *derivative time*.

采用 PID 控制器解决了许多实际控制问题。在标准的"教科书"中可以用方程:

$$u(t) = K\left(e(t) + \frac{1}{T_i}\int_0^t e(s)ds + T_d\frac{de(t)}{dt}\right)$$

描述 PID 控制器,其中 e 是指令信号 u_c(设定值)和过程输出 y(实测变量)之间的差。K 是增益即控制器的比例增益,T_i 是积分时间或重调时间,T_d 是微分时间。

第八节　直接数字控制算法
Section 8　Direct Digital Control Algorithm

并行程序设计法　parallel program design method
波纹　intersample ripple/hidden oscillation
采样开关　sampling switch
采样频率　sampling frequency
采样数据系统　sampled-data system
差分方程　difference equation
超调　overshoot
初值定理　initial-value theorem
串行程序设计法　serial program design method
存滞后　dead time
大林算法　Dahlin algorithm
单位加速度信号　unit acceleration signal
单位阶跃信号　unit step signal
单位速度信号　unit velocity signal
二阶惯性环节　second-order lag process
高阶保持器　high-order hold
广义对象　general object
控制规律　control law
离散系统　discrete-time system
零点和极点　zero and pole
零阶保持器　zero-order hold
脉冲传递函数　impulse transfer function
时间常数　time constant
输出幅度　output amplitude
数据采集系统　data acquisition system
数字控制器　digital controller
随动系统　follow system
响应曲线　response curve
一阶保持器　first-order hold
一阶惯性环节　first-order lag process
振铃现象　ringing phenomenon
直接程序设计法　direct program design method
终值定理　final-value theorem
自动跟踪　self tracking
最少拍控制　deadbeat control

1 This strategy has the property that it will drive all the states to zero in at most n steps after an impulse disturbance in the process state. The control strategy is called deadbeat control.

当系统受到脉冲扰动时,使系统达到稳定时所需要的采样周期最少,这种控制策略称为最少拍控制。

2 A special type of controller, called a deadbeat controller, responds quickly to the control input and settles out with zero error in minimum finite time. A deadbeat controller can be designed in the z-domain by replacing the closed-loop poles of the system with poles at the origin of the z-domain.

被称为一类特殊控制器的最少拍控制器,对控制输入响应快速,且输出在最短的有限时间内达到稳定。最少拍控制器能在 z 域内设计,用 z 域原点处的极点来替代系统的闭环极点。

第十二章　计算机控制技术

3 This type of controller design is actually a special case of the direct design. Deadbeat controllers, then, suffer from all the problems and constraints that need consideration when doing direct design. In addition, the controllers resulting from this design procedure tend to be highly sensitive to parameter variations within the system.

此类控制器设计，实际上是直接设计的一个特例。而在进行直接设计时，最少拍控制器要考虑各种问题和约束。此外，按此程序设计的控制器可能对系统内部的参数变化非常敏感。

4 Deadbeat design presents an optimal kind of design (optimal in the sense of minimal settling time) and is often useful for comparison purposes.

最少拍设计是一类最优的设计（最小过渡过程时间意义上的最优），通常用于比较的目的。

第九节　模糊控制技术
Section 9 Fuzzy Control Technology

单变量系统　single variable system
单输入单输出　single-input and single-output
多变量模糊控制　multi-variable fuzzy control
多变量系统　multi-variable system
多输入单输出　multi-input and single-output
非线性复杂系统　nonlinear complex system
经典控制理论　classical control theory
控制规则　control rule
模糊关系　fuzzy relation
模糊控制　fuzzy control
模糊控制算法　fuzzy control algorithm
模糊数学模型　fuzzy mathematics model
模糊条件语句　fuzzy condition clause
模糊语言　fuzzy language
神经网络　neuron network
数学模型　mathematics model
双输入单输出　double-input and single-output
现代控制理论　modern control theory
语言变量　language variable
专家模糊控制　expert fuzzy control
自适应控制　adaptive control
自学习　self learning

1 Fuzzy logic is a way of interfacing inherently analog processes, which move through a continuous range of values, to a digital computer to perform tasks, based on abstracted values, as if they were well-defined discrete numeric values.

模糊逻辑是基于抽象的值，将一定范围内变化的固有模拟过程和数字计算机进行接口执行任务，就好像它们是定义好的数字值一样。

2 The input variables in a fuzzy control system are in general mapped into by sets of membership functions similar to this, known as "fuzzy sets". The process of converting a crisp input value to a fuzzy value is called "fuzzification".

一般将模糊控制系统的输入变量映射成与之相似的成员函数的集合，即模糊集。将输入值变成模糊值的过程称为模糊化。

3 The fuzzy controller is composed of the following four elements:
(1) A rule-base (a set of IF-THEN rules), which contains a fuzzy logic quantification of the expert's linguistic description of how to achieve good control.
(2) An inference mechanism (also called an "inference engine" or "fuzzy inference" module), which emulates the expert's decision making in interpreting and applying knowledge about how to best control the plant.
(3) A fuzzification interface, which converts controller inputs into information that the inference mechanism can easily use to activate and apply rules.
(4) A defuzzification interface, which converts the conclusions of the inference mechanism into actual inputs for the process.

模糊控制器由下面四部分组成：
(1) 规则库（如果那么规则集），包含取得良好控制的专家语言描述的模糊逻辑量化。
(2) 推理机构（又称推理机或模糊推理模块），模仿专家制定决策，解释和应用知识以便更好地控制设备。

(3) 模糊化接口，将控制器的输入转换为推理机构易于应用的信息，激活或者应用规则。
(4) 去模糊化接口，将推理机构的结论转换为过程的实际输入。

第十节 微型计算机控制系统设计
Section 10 Microcomputer Controlled System Design

程序流程图	program flow chart	软件调试	software debugging
单片机系统	single chip system	软件开发过程	software development process
调试及实验	debugging and testing	系统设计	system design
工业控制机	industrial process computer	硬件调试	hardware debugging
控制方案	control scheme	原理图	principle map
离线	off-line	在线	on-line

1 The design of control system is a specific example of engineering design. Again, the goal of control engineering design is to obtain the configuration, specifications, and identification of the key parameters of a proposed system to meet an actual need.

控制系统设计是工程设计的特定例子。也可以说，控制工程设计的目标是获得被提议系统关键参数的结构、规格及标准，以满足实际的需要。

2 The first step in the design process is to establish the system goals. For example, we may state that our goal is to control the velocity of a motor accurately. The second step is to identify the variables we desire to control (for example, the velocity of the motor). The third step is to write the specification in terms of the accuracy we must attain. This required accuracy of control will then lead to the identification of a sensor to measure the controlled variable.

设计过程的第一步是建立系统的目标。如可以声明，控制的目标是精确地控制电机的速度。第二步是确定期望控制的变量(如电机的速度)。第三步写出关于必须达到的精度规格。那么依据要求的控制精度，就可以确定传感器的设计标准以测量被调变量。

第十三章 精密机械与仪器
Chapter 13 Precise Machinery and Instrument

第一节 绪论
Section 1 Introduction

安排(装置)　arrangement
构件　member n.
机构　mechanism
机器　machine
机械　machinery
机械的　mechanical
机械动力装置　mechanical power

机械学　mechanics
机械原理　mechanism theory
机械装置　mechanical device
精密机械　precise machinery
零件　component
仪器　instrument

1 Few products are as complex to develop and produce as gasoline-powered automobiles, which are assembled with thousands of precise machinery components.
燃油汽车使用成千上万的精密机械部件组装而成,几乎没有多少产品像开发和生产燃油汽车那么复杂。

2 Toyota said mechanisms in the gas pedal could become worn and make it harder to depress the accelerator and, in the worst case, allow the pedal to get stuck in a partially depressed position.
丰田称,这些汽车油门踏板处的机械装置有可能受到磨损,从而加大其下压汽车油门的难度,在最坏情况下,可能出现油门踏板踩到一半就卡壳的现象。

第二节 机械工程常用材料及钢的热处理
Section 2 Mechanical Project's Common Materials and Steel Heat Treatment

表面热处理　surface heat treatment
淬火　quenching
弹性极限　elastic limit
电镀　electroplating
非金属材料　non-metal material
刚度　rigidity
工程塑料　engineering plastic
合金钢　alloy steel
化学热处理　chemical heat treatment
灰铸铁　graphitic cast iron
回火　tempering
金属材料　metal material
抗拉强度　tensile strength

铝合金　aluminum alloy
疲劳极限　fatigue limit
屈服极限　yielding limit
屈服强度　yielding strength
塑性变形　plastic deformation
钛合金　titanium alloy
碳钢　carbon steel
铜合金　copper alloy
退火　annealing
橡胶　rubber
应力极限　stress limit
正火　normalizing

1 Non-metal materials are those materials such as woods, plastics, rubbers, epoxy resins, ceramics and diamonds that do not have a metallic base.
非金属材料是指如木头、塑料、橡胶、环氧树脂、陶瓷及金刚石等那些不含有金属基材的材料。

2 Plastic deformation is the ability to bend or deform before breaking.
塑性变形就是在断裂前弯曲或变形的能力。

3 Rigidity refers to the capacity to resist deformation of a component or an element.
刚度是指材料在外载荷作用下抵抗弹性变形的能力。

第三节 平面机构的结构分析
Section 3　Structure Analysis of Planar Mechanisms

闭式链　closed chain	球面副　globular pair
低副　lower pair	虚约束　formal constraint
分析　analysis	移动副　sliding pair
复合铰链　multiple pin joint	圆柱副　cylindric pair
高副　higher pair	约束　constraint
机构运动简图　kinematic scheme	运动副　joint / pair
机架　frame	运动副元素　element of kinematic pair
简图　scheme	运动确定性　kinematic determination
简写符号　abbreviation	转动副　rotating pair
结构分析　structure analysis	装配　assemble
局部自由度　partial freedom	自由度　freedom
开式链　open chain	自由度数　degree of freedom
链系　chain	组合　assemblage
平面副　flat pair	组合件　assembly
平面机构　planar mechanism	

1 The test results show that the wear of the pump in a hydraulic system mainly depends on the relation between the size of abrasive particles and the dynamic clearance dimension of the key moving parts.
结果表明,在液压系统中,泵的磨损主要决定于磨粒与泵内关键运动副动态间隙之间的尺寸关系。

2 Analysis and discussion were made upon the selection of the most important graphic element in plotting system of kinematic scheme designed by the authors.
作者分析讨论了设计的机构运动简图绘图系统中最重要的图元的选择。

3 Based on the analysis of the kinematic scheme of mechanism concerning the eccentric piston pump, the three-dimension model of the eccentric piston pump was established by CAXA solid design software.
在分析偏心柱塞泵机构运动简图的基础上,运用CAXA实体设计软件创建了该机构的三维模型。

第四节 平面连杆机构
Section 4　Planar Linkages

传动角　transmission angle	铰链四杆机构　four-bar linkage
传动特性　transmission characteristic	空回行程　return stroke
导杆　guidebar	空间连杆机构　spatial linkage
等腰双曲柄机构　isosceles linkage	连杆　coupler
对心曲柄滑块机构　central(radial) slider-crank mechanism	连杆曲线　coupler curve
	偏置曲柄滑块机构　eccentric slider-crank mechanism
反平行双曲柄机构　antiparallel crank linkage / (crossed parallelogram linkage)	平面铰链四杆机构　planar four-revolute linkage
工作行程　working stroke	
滑块　slider	平面连杆机构　planar linkage
滑块的导路　slider guide	平行四边形机构　parallelogram linkage
急回特性　quick-return characteristic	曲柄　crank
急回特性机构　quick-return mechanism	曲柄摆动导杆机构　crank-and-oscillating guidebar mechanism
急回运动　quick-return motion	
间歇运动的连杆机构　dwell linkage	曲柄存在条件　conditions of crank existence

第十三章　精密机械与仪器

曲柄导杆机构　crank-and-guidebar mechanism
曲柄滑块机构　slider-crank mechanism
曲柄摇杆机构　crank-and-rocker mechanism
曲柄摇块机构　crank-swing block mechanism
曲柄移动导杆机构
crank-and-translation guidebar mechanism
曲柄转动导杆机构
crank-and-rotating guidebar mechanism
十字滑块机构　crossed-slider mechanism
双滑块机构　double-slider mechanism
双曲柄机构　double-crank mechanism
双摇杆机构　double-rocker mechanism
双转动导杆机构
double rotating guidebar mechanism
死点　dead point
行程速度变化系数
coefficient of travel speed variation
摇杆　rocker
摇块机构　rocking-block mechanism
正弦机构　sine mechanism

1 The transmission angle is defined as the angle between the output link and the coupler. It is usually taken as the absolute value of the acute angle of the pair of angles at the intersection of the two links and varies continuously from some minimum to some maximum value as the linkage goes through its range of motion.
传动角是指输出杆和连杆间的夹角,它的值等于输出杆与连杆间所夹锐角,或等于所夹钝角的补角的值。它随机构的运动而不断在最大与最小值之间变化。

2 Worm gear increases the role of body twist. The output shaft drives four-bar linkage, through the four-bar linkage to change the continuous rotary motion from side to side movement.
蜗轮蜗杆机构的作用是(减速)增扭,其输出轴带动四连杆机构,通过四连杆机构把连续的旋转运动改变为左右摆动的运动。

第五节　凸轮机构
Section 5　Cams

摆动从动件　pivoted follower
摆线运动规律　law of cycloidal motion
从动件　follower member
从动件滚子　follower roller
等加速-等减速运动规律
law of constant acceleration and deceleration motion
等速运动规律　law of uniform motion
定宽凸轮　box cam
端面凸轮　bell cam
对心从动件　radial follower
刚性冲击　rigid impulse
滚子从动件　roller follower
基圆　base circle
尖端从动件　tip follower
简谐运动规律　law of simple harmonic motion
空间凸轮　spatial cam
盘形凸轮　disk cam /radial cam
偏置从动件　offset follower
平底从动件　plain-faced follower
球面凸轮　spherical cam
柔性冲击　soft impulse
升程　advanced travel
凸轮廓形　cam shape
凸轮理论廓线　theoretical cam profile
凸轮轮廓　cam profile
凸轮轮廓的尖点　sharp point at the cam profile
移动从动件　translating follower
移动凸轮　translating cam
余弦加速度运动规律　law of cosine acceleration motion
圆柱凸轮　barrel cam/cylindrical cam
圆锥形凸轮　conical cam
运动失真　distortion of motion
正弦加速度运动规律　law of sine acceleration motion

1 The definition of cam is the projection on a wheel or shaft, designed to change circular motion into up-and-down or back-and-forth motion.
凸轮是轮子或轴上的凸出物,它被用来使圆周运动转变为上下或前后运动。

2 Improve the work reliability and accuracy of the mechanism by optimizing the design of cam mechanism's motion law.

通过对凸轮机构运动规律的优化设计,提高机构的工作可靠性和工作精度。

3 Roller radius sizes of slave mechanism of cam machine affect the touch-stress of touch region and the thickness of lube film directly.
凸轮机构从动件滚子半径的大小直接影响凸轮副接触处的接触应力和润滑油膜厚度。

第六节　齿轮传动
Section 6　Gear Transmission Mechanism

阿基米德蜗杆　Archimedes worm
摆线齿轮　cycloidal gear
包络　envelope
背锥　back cone
变位齿轮　modified gear
标准齿顶高　standard addendum
标准齿高　standard depth
标准齿轮　standard gear
标准压力角　standard pressure angle
差动轮系　differential gear train
齿顶　crest
齿顶厚　crest width
齿顶压力角　pressure angle at tip circle
齿顶圆直径　diameter of addendum
齿顶圆半径　radius of addendum
齿根高系数　coefficient of addendum
齿根厚　addendum thickness
齿厚　circular thickness
齿廓工作段　active tooth contour
齿廓啮合基本定律　fundamental law of gear-tooth action
齿轮范成原理　principle of gear generating
齿轮系　train of gears
齿轮形插刀　gear-form generating cutter
齿面接触　contact of teeth flank
齿条形插刀　rack-form generating cutter
单级行星轮系　single planetary gear train
单头蜗杆　single-threaded worm
当量齿轮　virtual gear
顶隙　top clearance
顶隙系数　coefficient of top clearance
定轴轮系　ordinary gear train
多级行星轮系　compound planetary gear train
惰轮　idle gear
范成法　generating method
仿形法　copying method
非标准齿轮　nonstandard gear

工作齿廓　working tooth contour
公法线长度　base tangent length
滚齿机　hobbing machine
混合轮系　compound epicyclic gear train
基圆齿厚　base thickness
基圆直径　base diameter
极限啮合点　limiting contact point / (limiting meshing point)
渐开线齿轮　involute gear
渐开线函数　involute function(Inv)
接触线、啮合线　line of contact
径向间隙　bottom clearance
开始啮合点　beginning meshing point
理论啮合线　theoretical contact line
螺旋齿轮　spiral gear
啮合点　meshing point
啮合轨迹　meshing path
啮合角　angle of action
啮合平面　plane of action
啮合线　meshing line
全齿高　total depth
人字齿轮　double helical gear
实际啮合线长度　working length of meshing line
双头蜗杆　double-threaded worm
蜗杆　worm / worm screw
蜗杆与蜗轮　worm-and -wheel
蜗轮　worm gear/ worm wheel
蜗轮副　worm gear pair
蜗轮滚刀　worm gear hob
无齿侧间隙啮合方程式　equation of engagement with zero backlash
谐波齿轮传动　harmonic gear transmission
行星齿轮　planetary gear
行星轮系　planetary gearing / planetary gear train
行星转臂　planet cage
圆柱蜗杆　cylindrical worm

正常齿高制　system of full depth
正常全齿高　full depth
直齿圆锥齿轮　straight bevel gear
终止啮合点　end meshing point

周转轮系　epicyclic gear train
锥角　cone angle
锥距　cone distance

1 The definition of gear is a set of toothed wheels working together in a machine, e. p. such a set to connect a motor-car engine with the road wheels.
齿轮机构是指相互啮合的轮子，它们在机器中同时工作。如连接车轮和汽车发动机的齿轮装置。

2 The main component that make up an automatic transmission is planetary gear sets, which are the mechanical systems that provide the various, forward gear ratios as well as reverse.
自动变速器的主要组成部分是行星齿轮机构提供多个前进挡和倒挡的机械系统。

3 The epicyclic train is compoesd of a sun gear and a planet gear orbiting around the sun, and the planet gear is held in orbit by the arm.
轮系是使中心轮和行星轮绕固定轴线转动，且行星轮被转臂支承着。

第七节　带传动
Section 7　Belt Drives

V带传动　V-belt drive
包角　angle of contact
初拉力　initial tension
传动比　speed ratio/transmission ratio
传动装置　transmission/driving
从动带轮　driven pulley
带长　belt length
带传动　belt drive
导轮　idler pulley
多楔带　poly V-belt
滑动率　sliding speed
基准宽度　datum width
基准直径　datum diameter
减速比　speed reducing ratio
节宽　pitch width
紧边拉力　tight side tension

宽V带　wide V-belt
离心拉力　centrifugal tension
联组V带　joined V-belt
摩擦轮传动　friction wheel drive
平带传动　flat belt drive V
普通V带　classical V-belt
普通平带　conventional belt
松边拉力　slack side tension
同步带传动　synchronous belt drive
无级变速　infinitely variable speed
有效拉力　effective tension
圆带传动　round belt drive
增速比　speed increasing ratio
窄V带　narrow V-belt
张紧轮　tension pulley
主动带轮　driving pulley

1 Belts and chains represent the major types of flexible power transmission elements.
在柔性传动零件中，主要是皮带传动和链条传动。

2 It is an intrinsic physical phenomenon of friction tape transfer for elastic slip, which leads to unstable movements for the instantaneous transmission ratio and a low mechanical efficiency.
弹性滑动是摩擦型带传动固有的物理现象，引起带传动瞬时传动比不稳定及机械效率降低。

3 Synchronous belt is a kind of special belt which possesses the merits of gear, chain and belt transmissions. The outside material is used to protect the teeth and the skeleton material.
同步带是一种集齿轮传动、链传动、带传动的优点于一身的传动带，其外部材料(包覆布)是保护其齿部和骨架线绳的材料。

第八节 轴、联轴器、离合器
Section 8　Shaft, Coupling and Clutch

安全离合器　safety clutch
安全销　safety pin
半圆键　woodruff key
波纹管联轴器　coupling with corrugated pipe
槽销　grooved pin
超越离合器　overrunning clutch
齿式联轴器　gear coupling
单向离合器　one-way clutch
弹性挡圈　snap ring
弹性联轴器　resilient shaft coupling
弹性套柱销联轴器　pin coupling with elastic sleeves
弹性圆柱销　spring-type straight pin / spring pin
导向平键　feather key/dive key
电磁离合器　electromagnetic clutch
刚性联轴器　rigid coupling
钩头楔键　gib-head taper key
滚柱离合器　roller clutch
花键　spline
滑键　feather key
滑块联轴器　NZ clao type coupling
机械离合器　mechanically controlled clutch
夹壳联轴器　split coupling
渐开线花键　involute spline
键　key
键槽　key way
矩形花键　rectangle spline
开口销　cotter pin/split pin
开尾圆锥销　taper pin with split
空心销轴　hollow pin
离合器　clutch
离心离合器　centrifugal clutch
联轴器　coupling
链条联轴器　chain coupling
轮胎式联轴器　coupling with rubber type element
梅花形弹性联轴器　coupling with elastic spider
摩擦式离合器　friction clutch
平键　flat key
普通平键　general flat key
切向键　tangential key
球笼式同步万向联轴器　synchronizing universal coupling with ball and sacker
蛇形弹簧联轴器　serpentine steel flex coupling
十字滑块联轴器　oldham coupling
实心销轴　solid pin
双作用离合器　dual clutch
套筒联轴器　sleeve coupling
凸缘联轴器　flange coupling
万向联轴器　universal joint
橡胶板联轴器　coupling with rubber plates
销　pin
销轴　clevis pin with head
楔键　taper key
牙嵌式离合器　jaw clutch
牙嵌式联轴器　jaw and toothed coupling
圆柱销　cylindrical pin/straight pin
圆锥销　conical pin/taper pin
轴端挡圈　lock ring at the end of shaft
轴肩挡圈　ring for shoulder
自控离合器　auto-controlled clutch

1 In rotating machinery system, the flexible coupling not only plays a part in transmitting the torque, but also in reducing the torsional impact.
在旋转机械系统中,柔性联轴器不但起传递转矩的作用,而且还有减小扭转冲击的作用。

2 By mounting overrunning clutch on the pumping unit, the interference of the motor on the balance of the unit can be avoided, and flywheel effect of the counterbalance weight can be fully achieved.
在抽油机适当部位加装超越离合器可解决抽油机平衡运动受电动机干扰的问题,能充分发挥平衡块的飞轮效应。

第九节 支承
Section 9　Bearing

保持架　cage
背对背配置(滚动轴承)　back-to-back arrangement (rolling bearing)
成对安装(滚动轴承)　paired mounting (rolling bearing)
尺寸系列(滚动轴承)

dimension series (rolling bearing)
单列轴承　single row bearing
单向推力轴承　single direction thrust bearing
当量载荷　equivalent load
调心滚子轴承　self-aligning roller bearing
调心轴承　self-aligning bearing
动载荷　dynamic load
多油楔滑动轴承　multi-oil wedge bearing
额定寿命　rating life
防尘盖　shield
防尘盖轴承　shielded bearing
复合轴承材料　compost bearing material
滚动体　rolling element
滚动轴承　rolling bearing
滚针　needle roller
滚针轴承　needle roller bearing
滑动轴承　plain bearing/ sliding-contact bearing
基本额定寿命　basic rating life
角接触轴承　angular contact bearing
径向滑动轴承　radial sliding bearing
径向载荷系数　radial load factor
径向—止推滑动轴承　thrust-purnal bearing
静载荷　static load
宽度系列（滚动轴承）
width series (rolling bearing)
密封圈　seal
密封圈轴承　sealed bearing
面对面配置（滚动轴承）
face-to-face arrangement (rolling bearing)
磨合性　running-in ability
磨损度　wear intensity
内圈　inner ring
偏心率　relative eccentricity
剖分式滑动轴承　split plain bearing
嵌入性　embeddability
球轴承　ball bearing
深沟球轴承　deep groove ball bearing
寿命系数　life factor

双列轴承　double row bearing
双向推力轴承　double direction thrust bearing
推力轴承　thrust bearing
外圈　outer ring
外形尺寸（滚动轴承）
boundary dimension (rolling bearing)
向心滚子轴承　radial roller bearing
向心球轴承　redial bearing
液体动压滑动轴承　hydrodynamic bearing
液体静压滑动轴承　hydrostatic bearing
仪器精密轴承　instrument precision bearing
油槽　oil groove
油孔　oil hole
圆柱滚子轴承　cylindrical roller bearing
圆锥滚子轴承　tapered roller bearing
整体式滑动轴承　solid bearing
直径系列（滚动轴承）
diameter series (rolling bearing)
止推滑动轴承　plain thrust bearing
轴承　bearing
轴承衬　bearing liner
轴承承载能力　bearing load carrying capacity
轴承径向载荷　bearing radial load
轴承宽度　bearing width
轴承内径　bearing bore diameter
轴承套圈　bearing ring
轴承外径　bearing outside diameter
轴承系列（滚动轴承）
bearing series (rolling bearing)
轴承压强　bearing mean specific load
轴承轴向载荷　bearing axial load
轴颈　journal
轴套　bearing bush
轴瓦　liner
轴向载荷系数　axial load factor
锥孔轴承　tapered bore bearing
自动调心滑动轴承　plain self-aligning bearing
自润滑滑动轴承　self-lubricating bearing

1 The purpose of a bearing is to support a load while permitting motion between two elements of a machine.
轴承的作用是在机器负载的情况下使两个构件正常运动。

2 The components of ball bearings are four main parts. The ball bearing, the inner and outer rings providing the pathway on which the balls move, the retainer or cage maintains separation of the balls, and it is a shield to keep dirt away from entering or a seal to retain lubrication.
球轴承有四个主要部分，滚珠、内环和外环——它们是为滚珠的移动设置轨道，还有保护架——它是为了将滚珠彼此隔开，它还可以防止灰尘进入和保留润滑油。

③ For parts such as gearwheels, sliding bearings and bushings, the wear on the part itself and on the counter materials must be restricted to a minimum.
对于齿轮、滑动轴承和轴衬这样的部件，对摩擦件本身和摩擦副材料的磨损要求降低到最小限度。

④ The advantage of rolling bearing is that they cause less friction.
滚动轴承的优点是它产生的摩擦力较小。

⑤ As standardized parts rolling bearing has a range of application in machineery, and its quality of installation has directly influenced the service life of the equipment.
滚动轴承是机械设备中应用较广的标准化部件，其安装质量的好坏，直接影响设备的使用寿命。

第十节　零件的精度设计与互换性

Section 10　Precision Design and Interchangeability of Machine Elements

标准公差　standard tolerance
标准公差等级　standard tolerance grade
标准化　standardization
表面粗糙度测量仪器　surface roughness measuring instrument
表面粗糙度的测量　surface roughness measuring
常用配合　commonly used fit
尺寸　size/dimension
尺寸公差　dimensional tolerance / size tolerance
垂直度　perpendicularity
单一要素　single essential factor
定位公差　location tolerance
定向公差　orientation tolerance
对称度　symmetry
公差带　tolerance zone
公差带代号　tolerance zone symbol
公差等级　tolerance grade
公差原则　tolerance principle
关联要素　associate essential factor
过渡配合　transition fit
过盈　interference
过盈配合　interference fit
互换性　interchangeability
基本尺寸　basic size
基本偏差　fundamental deviation
基孔制配合(ISO)　"hole-basis" system of fit
基轴制配合(ISO)　"shaft-basis" system of fit
基准孔　basic hole
基准制　benchmark system
基准轴　basic shaft
极限尺寸　limit of size
极限偏差　limit of deviation
间隙　clearance
间隙配合　clearance fit
局部实际尺寸　actual local size

孔　hole
理想要素　ideal feature
零线　zero line
轮廓算术平均偏差　Ra
轮廓中线　mean line
轮廓最大高度　Ry
面轮廓度　profile of any plane
配合　fit
配合表面　mating surface
配合公差　variation of fit / fit tolerance
配合公差带　fit tolerance zone
偏差　deviation
平面度　flatness
平行度　parallelism
评定长度　evaluation length
倾斜度　angularity
取样长度　sampling length
全跳动　total runout
上偏差　upper deviation
实际尺寸　actual size
实际偏差　actual deviation
实际要素　real feature / actual feature
跳动公差　run-out tolerance
同轴度　concentricity
微观不平度十点高度　Rz
位置度　position
位置公差　tolerance in position / position tolerance
下偏差　lower deviation
线轮廓度　profile of any line
形状公差　tolerance in form / form tolerance
优先配合　preferred fit
优先数系　series of preferred numbers
圆度　roundness / circularity
圆跳动　circular runout
圆柱度　cylindricity

第十三章 精密机械与仪器

直线度　straightness
轴　shaft
最大过盈　maximum interference
最大极限尺寸　maximum limit of size
最大间隙　maximum clearance

最大实体极限　maximum material limit (MML)
最小过盈　minimum interference
最小极限尺寸　minimum limit of size
最小间隙　minimum clearance
最小实体极限　least material limit (LML)

1 Tolerance of size and fit are the main contents of precision design.
尺寸公差与配合是精密设计时的主要内容。

2 The distribution of pre-stress in the multi-layer structure is related to the global and local rigidity, the contact area and the clearance of superimposition.
多层结构内部的预紧力的分布与结构的整体刚度、结合部位的局部刚度、接触面积、过盈配合间隙等因素有关。

3 For bearings having a tapered bore, inner rings are always mounted with an interference fit.
对于圆锥孔轴承，内圈的安装始终采用过盈配合。

4 Choosing form tolerance and position tolerance is an important content of the precision design.
确定形状公差和位置公差是精密设计的一个重要内容。

5 In the processing of dies, the technological process usually adopted for matching the guide bushes is the overfilling method, namely the pressing-in method.
模具加工过程中，固定导套通常采用的工艺是通过过盈配合将导套压入导套固定板的方法，即压入法。

6 The surface roughness is an index that judges of the parts surface quality.
表面粗糙度是衡量零件表面质量的一个指标。

7 In key coupling, the key width is benchmark, employing shaft-basic system of fits.
键联结中，键宽是基准，采用基轴制。

第十四章 智能楼宇
Chapter 14 Intelligent Building

第一节 绪论
Section 1 Introduction

安防自动化 security automation(SA)
办公自动化系统
office automation system (OAS)
综合布线系统 generic cabling system (GCS)
都市办公大楼 city place building
管理自动化 management automation(MA)
楼宇自动化系统
building automatic system(BAS)
消防自动化 fire automation(FA)

通信自动化系统
communication automatic system (CAS)
虚拟现实 virtual reality(VR)
异步传输模式
asynchronous transfer mode(ATM)
智能建筑物系统
intelligent building system(IBS)
智能楼宇 intelligent building

1 Since January 1984, the United States Connecticut State Hartford City built the first world-recognized intelligent buildings "Town Plaza", smart buildings swept the globe.
自1984年1月,美国康涅狄克州哈特福德市建成世界公认的第一座智能化大厦"城市广场"以来,智能建筑风靡全球。

2 Intelligent buildings are bound to be the product of IT era and the integration of high-tech with modern architecture, and it is a gigantic system project.
智能建筑是信息时代的必然产物,是高科技与现代建筑的有机结合,是一项巨大的系统工程。

第二节 楼宇自动化控制技术基础
Section 2 Fundamentals of Building Automation Control Technique

表示层 presentation layer
过程现场总线 process fieldbus (Profibus)
会话层 session layer
霍尼韦尔 Honeywell
集散型控制系统 distributed control system (DCS)
江森自控 Johnson controls
可寻址远程传感器数据协议 highway addressable remote transducer protocol (HART)
控制局域网 control area network (CAN)
流量传感器 flow sensor
设备描述 device description (DD)
设备描述语言 device description language (DDL)

湿度传感器 humidity sensor
实时数据库 real time database (RTDB)
数据链路层 data link layer
网络层 network layer
温度传感器 temperature sensor
物理层 physical layer
西门子 SIEMENS
现场总线控制系统 fieldbus control system (FCS)
压力传感器 pressure transducer
液位传感器 fluid level sensor
应用层 application layer
直接数字控制器 direct digital controller (DDC)

1 Building Automation System can help facility managers optimize energy, operations and indoor comfort over the entire lifetime of the green building and remain green in green building.
楼宇自动化控制系统可以节约能源,保证建筑正常运行,不断提供舒适的生活工作环境,并且使绿色建筑保持常青。

2 The building automatic control system is the main branch of the control system and the network is the master technological method of the building automation.
楼宇自动控制系统是控制系统的主要分支,网络是楼宇自动化的主要技术手段。

3 Now Ethernet has already been widely used in administration layer and supervision layer of BAS.

With the development of its own technology, Ethernet is being extended downwards to on-the-spot equipment of the intelligent buildings, applied directly to the communication among the on-the-spot equipment such as transformers, executors, DDCs, etc. The whole network is unified on Ethernet, from administration layer, supervision layer to field equipment layer.

现在以太网技术已经在楼宇自控系统的管理层、监控层得到广泛应用。随着其自身技术的发展，以太网技术正积极地向下延伸，应用于智能建筑的控制现场，直接应用于变送器、执行机构、DDC等现场设备间的通信，实现了从管理层、监控层到现场设备层的基于以太网的统一。

第三节 楼宇设备自动化系统
Section 3 Building Automation System

变风量　vary air volume(VAV)
点对点协议　point-to-point protocol(PPP)
电力系统　power system
电梯控制　lift control
电网　power network
调速电机　adjustable-speed motor
多点控制单元　multipoint control unit(MCU)
分散控制器　distributed control panel(DCP)
风机盘管　fan coil unit
负荷　load
给排水设备　plumbing equipment
给排水系统　plumbing system
空调系统　air-conditioning system
空气处理机　air handling unit(AHU)
联锁操作　interlocked operation
流量开关　flow rate switch
脉冲输入　pulse input(PI)
模/数　analog to digital(A/D)
模拟量输出　analogue output(AO)
模拟量输入　analogue input(AI)
暖通空调　heating ventilation air conditioning (HVAC)
配电系统　electric distribution system
普通型分散控制器　general distributed control panel
群控电梯　multiple-control lift
数/模　digital to analog(D/A)
数据采集器　data gathering panel
数据库系统　database system
数字量输出　digital output(DO)
数字量输入　digital input(DI)
送风机　blower
网络接口单元　network interface unit(NIU)
网络控制系统　network control system(NCS)
网络终端设备　network terminal equipment(NTE)
照明控制器　lighting controller
照明系统　illuminator
智能型分散控制器　intelligent distributed control panel

1 The effective use of lighting can have significant energy saving effect in buildings. The survey shows that the reason for the less importance of lighting system in intelligent buildings is probably due to the fact that the respondents consider the maximization of daylight resource has the potential to improve the quality of indoor lighting and substantially reduce the consumption of artificial lighting as well as energy costs.

有效地利用灯光能够显著减少大楼的能量使用。调查者认为照明系统在智能建筑中的作用较小，可能是因为：如果将日光引进房间且最大限度地利用，能够提高室内光照的效果，并且能够大大减少人工照明的能量消耗以及能源费用。

3 While there are a number of intelligent building components or products available in the market, decision makers are confronted with the task of forming a particular combination of components and products to suit the need of a specific intelligent building project (for example, building automation system, HVAC, lighting, electrical installation, lift, fire protection, safety and security system), and simultaneously resolving any conflicts between the performance criteria.

尽管市场上提供了大量的智能建筑组件或产品，但决策者们仍然面临着选出一套恰当的组件或产品组合，以适应特种智能建筑项目的需要，(比如建筑自动化系统、HVAC、照明、电器安装、电梯、防火及安保系统)，同时解决这些性能标准之间的冲突问题。

第四节　火灾自动报警与控制
Section 4　Fire Automation Alarm and Control

备用电源　standby power	监控台　control and monitor console
多线制　multiple-way system	监控系统　monitored control system
防火安全门　fire door	控制模块　control module
防火幕　fire curtain	喷洒灭火系统　sprinkler system
公共广播　public address(PA)	消防给水　fire-water supply
公共广播系统　public address system(PAS)	消防设备　fire-fighting equipment
光/电转换器　optical to electrical converter	消防栓　fireplug
火警系统　fire alarm system	消防用水　fire demand
火灾管理系统　fire management system	消防自动化系统　fire-fighting automation system
火灾红外探测器　infrared flame-failure detector	烟雾报警器　smoke alarm
火灾控制器　fire controller	阴燃　smoulder
监控设备　monitoring equipment	中央控制室　central control room
监控室　supervisory control	

1 Fire protection in intelligent building is critical as it can contribute significantly to the success of rescue operations and to minimize the degree of damage.
防火措施在智能建筑中也是至关重要的，因为它可以确保及时营救伤者，并且将损伤程度降至最低。

2 The traditional automatic fire alarm system no longer fulfills the need of large scattered industry constructions. Therefore automatic networked fire alarm systems appeared.
传统的火灾自动报警系统已经不适应这种分散的大规模工业建筑的需要，因此出现了网络型火灾自动报警系统。

3 Although the application of fire signal processing algorithms has brought benefit to the reduction of false alarm, the improvement and development of fire signal processing algorithms is still a main research subject in fire detection field.
火灾探测信号处理算法的应用促进了火灾探测技术的进步，减少火灾探测中的误报问题。但是，如何提高检验火灾探测信号处理算法的性能，仍是一个重要的研究课题。

4 In recent years, fire alarm system based on machine vision is brought forward; the system realize fire auto alarm using digital image processing technique.
近年来提出的基于机器视觉的火灾报警系统，利用数字图像处理技术来实现火灾自动报警。

第五节　楼宇安全防范技术
Section 5　Building Security Automation Techniques

IC卡　integrated circuit card	电子锁　electronic lock
报警传感器　alarm sensor	读卡机　card reader
报警系统　alarm system	个人识别号　personal identification number(PIN)
报警装置　alarm device	红外线探测器　infrared director
被动式红外线传感　passive infrared detector(PID)	门禁　control at gate
闭路电视监视系统　closed circuit television(CCTV)	智能建筑管理系统
出入口控制系统　access control system	intelligent building management system(IBMS)
磁卡　magnetic card	智能卡　smart card
有线电视　cable television(CATV)	智能控制器　intelligent controller

1 Access control system, which belongs to building automation system (BAS), can not only control entranceways of building, but also combine with the fire fighting and alarming system and closed-circuit television system.

门禁保安系统是楼宇自动化系统的一个子系统,它能够完成对楼宇的出入口的控制以及与消防报警系统和闭路监视系统的联动控制。

2 Now in intelligent residential area, the safe precautionary technology is one of the most important technology.

在当今的智能小区技术中,安全防范技术是其中的一项重要技术。

3 According to the design, the family's smart terminal system can carry out the family defence, synchronous action, distant surveillance and control by telephone or Internet, family amusement, information service, etc..

根据设计,该家庭智能终端系统可以实现安全防范、联动控制、电话远程监视与控制网络远程监视与控制、家庭娱乐、小区信息服务等功能。

4 The idea of a smart home might make you think of George Jetson and his futuristic abode or maybe Bill Gates, who spent more than 100 million dollars building his smart home.

智能家庭的概念可能会使你联想到乔治·杰特森和他未来派的住所,或是比尔·盖茨那栋耗费一亿多美元建造的智能家居。

5 For example, not only would a resident be woken with notification of a fire alarm, the smart home would also unlock doors, dial the fire department and light the path to safety.

比方说家里突然起火了,智能家居不仅会自动报警,而且会给你打开房门、自动向消防队拨打电话,并照亮你逃生的路途。

第六节 综合布线技术
Section 6　Generic Cabling Techniques

表面传输阻抗　surface transfer impedance(STI)
电子数据交换　electronic data interchange(EDI)
多用户多媒体插座
multi-user multimedia outlet(MMO)
多用户信息插座
multi-user information outlet(MIO)
非屏蔽双绞线　unshielded twisted pair(UTP)
分配线架　intermediate distribution frame(IDF)
光纤到家庭　fiber to the home(FTTH)
光纤到桌面　fiber to the desk(FTTD)
光纤互连　fiber interconnection unit(FIU)
光纤同轴电缆混合系统　hybrid fiber coax(HFC)
会议电视系统　video conference system(VCS)
建筑群配线架
campus distributor(CD)
建筑物布线系统
premises distribution system(PDS)
建筑物配线架　building distributor(BD)
建筑物智能系统集成　building intelligent system integration(BISI)
交互式电视　interactive television(ITV)
结构化布线系统　structured cabling system(SCS)
介质接口连接器
media interface connector(MIC)
金属箔双绞电缆　foil twisted pair(FTP)

决策支持系统　decision support system(DSS)
卡口式光纤连接器
bayonet fiber optic connector(BFOC)
楼层配线架　floor distributor(FD)
模块化楼宇控制器
modular building controller(MBC)
模块化设备控制器
modular equipment controller(MEC)
屏蔽双绞线　shielded twisted pair(STP)
企业建筑物集成
enterprise buildings integrator(EBI)
设备间　equipment room(ER)
视频点播　video on demand(VOD)
数据通信适配器
digital communication adapter(DCA)
数字用户单元　digital line unit(DLU)
双工 SC 连接器　duplex SC connector(SC-D)
通信插座　telecommunication outlet (TO)
通信网络　communication network system(CNS)
信息插座　information outlet(IO)
用户连接器　subscriber connector(SC)
直通式光纤连接器　straight tip(ST)
主配线架　main distribution frame(MDF)
住宅产业集成建造系统
housing industrialization contemporary integrate

manufacture system(HI-CIMS)
转接点　transition point (TP)
桌面型会议电视系统
desktop video conference system(DVCS)
综合布线　generic cabling(GC)
综合布线系统　generic cabling system (GCS)
综合业务数据网
integrated services digital network(ISDN)
综合语音数据终端
integrated voice data terminals(IVDT)

1 The integrated wire laying, safety and protection system, modeling and budgeting of an intelligent community play an important role in the research of the whole system.

智能小区的综合布线、安全防范系统及其选型和预算在整个系统的研究上都起着重要的作用。

2 The intelligent building automation systems and intelligent building cabling system is an important part of intelligent building. It is related to intelligent buildings and intelligent level.

智能建筑自动化系统与综合布线系统是智能建筑的主要组成部分,它关系到智能建筑的智能化水平。

3 Cabling products are mainly used to connect PC, computer network facilities and shared equipment. Besides its close relationship with the network facilities, cabling industry in China has grown up as an independent market with its unique history and characteristics.

综合布线产品主要用于计算机网络中各种PC、网络设备和外设的连接。而中国大陆地区的综合布线产业既与网络产品密切联系,又形成了一个较为独立的市场,有其独特的发展历史和特性。

第十五章 无线传感器系统
Chapter 15 Wireless Sensor System

第一节 概述
Section 1 Summary

传输层	transport layer
低复杂度	low complicacy
低功耗	low-power
低数据速率	low data rate
低延迟	low-latency
多跳路由	multi-hop routes
工作标准	standardization efforts
开源操作系统	open-source operating system
数据链路层	data link layer
体系结构	architecture
网络层	network layer
网状网络	mesh network
微机电系统	micro electro-mechanical systems
无线传感器	wireless sensor
无线个域网	wireless personal area networks
物理层	physical layer
协议栈	protocol stack
应用层	application layer
应用程序框架	application framework
自组织	self-organization

1 WSNs are composed of a large number of sensor nodes, which are densely deployed either inside a physical phenomenon or very close to it. In order to enable reliable and efficient observation and to initiate the right actions, physical features of the phenomenon should be reliably detected/estimated from the collective information provided by the sensor nodes.

无线传感器网络是由大量部署在被测物体内部或周围的传感器节点组成,这些节点用来检测被测物理量的实时数据或者进行数据预判,通过这些数据可以对该被测量进行准确、有效的观察,进而为下一步正确执行动作提供依据。

2 WSNs may consist of many different types of sensors including seismic, magnetic, thermal, visual, infrared, acoustic, and radar, which are able to monitor a wide variety of ambient conditions that include the following: temperature, humidity, pressure, speed, direction, movement, light, soil makeup, noise levels, the presence or absence of certain kinds of objects, and mechanical stress levels on attached objects. As a result, a wide range of applications are possible. This spectrum of applications includes homeland security, monitoring of space assets for potential and human-made threats in space, ground-based monitoring of both land and water, intelligence gathering for defense, environmental monitoring, urban warfare, weather and climate analysis and prediction, battlefield monitoring and surveillance, monitoring of seismic acceleration, strain, temperature, wind speed.

无线传感器网络中可以包括震动传感器、电磁传感器、热量传感器、视觉传感器、红外传感器、声音传感器、雷达等多种类型的传感器。这些传感器的运用可以用来对温度、湿度、压力、速度、方向、运动、光照、土壤成分、噪音水平、物体缺少物质和附着物的机械应力水平等进行监测。因此,无线传感器网络可以在绝大部分应用场合中使用。这些应用场合包括:国土安全、监测潜在或者人为的外太空威胁、地面土地和水资源监测、国防情报搜集、环境监测、城市战、气候及天气的分析预测、战场形势监测、地震加速度、应变、温度、风速的监测。

第二节 初识 ZigBee
Section 2 Hello ZigBee

AA 电池	AA battery
ZigBee 规范	ZigBee specification
ZigBee 联盟	The ZigBee Alliance
加密	encryption
开放的全球标准	open global standard
蓝牙	Bluetooth
偏移正交相移键控	Offset-Quadrature Phase-Shift Keying(O-QPSK)

商业楼宇自动化　commercial building automation
工业厂房监控　industrial plant monitoring
射频识别技术
Radio Frequency Identification(RFID)
无线自组网按需平面距离矢量路由协议　Ad hoc On-Demand Distance Vector Routing（AODV）
信道监听　listens to the channel
载波侦听　多路访问冲突避免 Carrier Sense Multiple Access Collision Avoidance（CSMA-CA）
帧校验　frame checksum
直接序列扩频
Direct Sequence Spread Spectrum(DSSS)
智能家居　home automation
自愈环网　self-healing ring network
总线拓扑　bus topology

1 The ZigBee wireless networking standard fits into a market that is simply not filled by other wireless technologies. While most wireless standards are striving to go faster, ZigBee aims for low data rates. While other wireless protocols add more and more features, ZigBee aims for a tiny stack that fits on 8-bit microcontrollers. While other wireless technologies look to provide the last mile to the Internet or deliver streaming high-definition media, ZigBee looks to control a light or send temperature data to a thermostat. While other wireless technologies are designed to run for hours or perhaps days on batteries, ZigBee is designed to run for years. And while other wireless technologies provide 12 to 24 months of shelf life for a product, ZigBee products can typically provide decades or more of use.

ZigBee 无线网络标准在市场占有其他无线技术不可替代的地位,原因主要在于：当市面上大多数无线标准一味地提高传输速率时,ZigBee却专注于低速率传输。当其他的无线协议栈追求添加更多的特性时,ZigBee协议栈却将注意力集中在8位微控制器适用的微型协议栈。当其他无线技术更多地关注互联网最后一英里的解决方案或者提供更高清晰度的流媒体的时候,ZigBee却更专注于如何去控制一盏灯或者如何将温度数据送至恒温控制器。当其他的无线技术在研究如何使自己的产品在使用电池供电的情况下运行几小时或者几天的时候,ZigBee却在考虑如何使其产品能够运行几年。当其他的无线技术在产品可靠性上致力于使自己产品能够连续一年或者两年无故障时,ZigBee却想着如何使自己的产品在十年或者更长的时间内都是可靠的。

2 ZigBee does not exactly fit the OSI 7-layer networking model, but it does have some of the same elements, including the PHY (physical), MAC (link layer), and NWK (network) layers. Layers 4 - 7 (transport, session, presentation, and application) are wrapped up in the APS and ZDO layers in the ZigBee model.

ZigBee 不完全适用于 OSI 7 层网络模型,但是其中有一些相同的因素存在,例如：物理层、数据链路层、网络层。第4-7层（传输层、会话层、表示层、应用层）被包含在了ZigBee模型中的APS层和ZDO层。

第三节　ZigBee 应用
Section 3　ZigBee Applications

MAC 地址　MAC Address
簇　cluster
单点传送　unicast
端点　endpoint
多点传送　multicast
公共规范　public profiles
广播　broadcast
间接寻址　indirect addressing
简化功能设备　reduced functional device (RFD)
链路质量指示　link quality indicator
路由器　router
能量检测　energy detection
请求　request
确认　confirm
设备类型　device types
数据请求　data request
特定规范　manufacturer-specific profiles
完整功能设备　full functional device(FFD)
网络地址　network address
响应　response
协调器　coordinator
协议栈规范　stack profile
信标　beacon
信道　channel
应用支持子层　Application Support Sublayer(APS)
原语　primitive

第十五章 无线传感器系统

直接寻址　direct addressing
指示　indication
终端设备　end-device
属性　attribute

1 This section describes PAN IDs, extended PAN IDs, and channels, all concepts which define a single ZigBee network. ZigBee PANs are formed by ZigBee Coordinators. Only ZigBee Coordinators (ZCs) may form a PAN. The other ZigBee node types, ZigBee Routers (ZRs) and ZigBee End-Devices (ZEDs) may join a network, but do not form one themselves.

这一章节描述了个域网编号、扩展的个域网编号和信道,这些概念定义了一个唯一的 ZigBee 网络。ZigBee 个域网是由 ZigBee 协调器创建,并且只能由其创建。其他类型的节点类型,如路由器和终端节点只能够加入该网络,但是并不能创建网络。

2 Application profiles are agreements for messages, message formats and processing actions that enable developers to create an interoperable, distributed application between applications that reside on separate devices. These application profiles enable applications to send commands, request data and process commands and requests. For instance, a thermostat on one node can communicate with a furnace on another node. Together, they cooperatively form a heating application profile. ZigBee vendors develop application profiles to provide solutions to specific technology needs.

应用规范是一组统一的消息、消息格式和处理方法,允许开发者建立一个可以共同使用的分布式应用程序,这些应用是利用独立设备中的应用实体来实现的。这些应用规范允许应用程序发送命令、请求数据、处理命令和请求。例如,一个节点上的恒温器可以与另外一个节点上的炉子通信,这两个节点协同工作就构成了一个温控的应用规范。ZigBee 技术服务商的工作就是根据不同的应用需求开发出与之相适应的应用规范。

3 The application framework in ZigBee is the environment in which application objects are hosted on ZigBee devices. Inside the application framework, the application objects send and receive data through the APSDE-SAP.

The data service, provided by APSDE-SAP, includes request, confirm, response and indication primitives for data transfer. The request primitive supports data transfers between peer application object entities. The confirm primitive reports the results of a request primitive call. The indication primitive is used to indicate the transfer of data from the APS to the destination application object entity.

Up to 240 distinct application objects can be defined, each interfacing on an endpoint indexed from 1 to 240. Two additional endpoints are defined for APSDE-SAP usage: endpoint 0 is reserved for the data interface to the ZDO and endpoint 255 is reserved for the data interface function to broadcast data to all application objects. Endpoints 241-254 are reserved for future use.

ZigBee 中的应用框架是为驻扎在 ZigBee 设备中的应用对象提供活动的环境。在应用框架中应用对象之间发送接收数据是通过应用支持层数据实体——服务通道(APSDE-SAP)来实现的。

由 APSDE-SAP 提供的数据服务包括数据传输过程中的请求、确认、响应和指示原语。请求原语支持应用对象实体之间的数据传输。确认原语用来报告请求原语的结果。指示原语用来表示从 APS 到目标应用程序对象实体之间的数据传送。

最多可以定义 240 个相对独立的应用程序对象,并且任何一个对象的端点编号都是从 1 到 240。此外还有两个附加的终端节点,为了 APSDE-SAP 的使用:端点号 0 固定用于 ZDO 数据接口;另外一个端点 255 固定用于所有应用对象广播数据的数据接口功能。端点 241-254 保留。

第四节 ZigBee、ZDO 和 ZDP
Section 4　ZigBee, ZDO, and ZDP

ZigBee 设备对象　ZigBee Device Object(ZDO)
ZigBee 设备描述　ZigBee Device Profile(ZDP)
绑定　binding
电源描述符　power descriptor
端点描述符　endpoint descriptor
端点匹配　matching endpoint

分片	fragmentation	设备发现	device discovery
服务发现	service discovery	设备描述符	device description
复杂描述符	complex descriptor	设备声明	device announce
简单描述符	simple descriptor	用户描述符	user descriptor
节点	node	源绑定	source binding
节点描述符	node descriptor	终端设备绑定	end device binding

1 This application, ZDO, keeps track of the state of the ZigBee device on and off the network, and provides an interface to the ZigBee Device Profile (ZDP), a specialized Application Profile (with profile ID 0x0000) for discovering, configuring, and maintaining ZigBee devices and services on the network.

ZDO not only interacts with APS, but also interacts directly with the network layer. ZDO controls the network layer, telling it when to form or join a network, and when to leave, and provides the application interface to network layer management services. For example, ZDO can be configured to continue attempting to join a network until it is successful, or until a user-specified number-of-retries has occurred before giving up, and informing the application of the join failure.

ZDO 用于持续追踪 ZigBee 设备开启、关闭网络的状态，为 ZigBee 设备描述(ZDP)提供接口，并且使用一个专门的应用规范(ID号 0x0000)用来发现、配置和维护网络上的 ZigBee 设备和服务。

ZDO 不仅与设备支持子层(APS)相连，它还直接与网络层进行通信。ZDO 对网络层的控制体现在告知它何时形成或者加入网络，何时离开网络，并且为网络层的管理服务提供应用程序接口。例如，ZDO 可以配置不断地尝试加入网络直到它加入成功，或者用户设定重试加入次数的上限，超过该上限则告知应用程序加入失败。

2 The ZigBee Device Profile (ZDP) contains a set of commands for discovering various aspects about nodes in the network. The ZigBee specification calls these "device discovery services", which can be confusing because endpoints contain device IDs which really describe individual ZigBee applications running in that node. So, when you see ZDP Device Discovery, think node-wide (not application/endpoint specific) services.

ZigBee 设备描述(ZDP)包含了一套命令，该命令用于发现网络节点的各种信息。ZigBee 规范称之为"设备发现服务"，令人困惑的是端点包含的设备 ID 才是真正的区分节点上 ZigBee 应用的标识。所以当遇到 ZDP 设备发现时，想一想节点层面的服务而不要考虑应用程序或者特定的端点层面。

3 ZigBee uses descriptors to describe a node and its properties, allowing other applications running in the network to discover these properties over-the-air. Node-wide descriptors include the node descriptor, the power descriptor, the complex descriptor, and the user descriptor.

The node descriptor contains a variety of fields, including the node type of the device (whether the node is a ZigBee Coordinator, Router, or End-Device), the manufacturer's code, whether the optional user and complex descriptors are present, and whether the node supports fragmentation.

The other descriptors include the power descriptor, which defines which power modes this node supports, and the user descriptor, which contains a user definable string to identify the location (such as living room or office). These descriptors are all optional in the ZigBee spec. The user descriptor is settable over-the-air, the rest are only gettable.

ZigBee 用描述符来描述该节点及其特性，并且允许网络中其他运行着的应用程序以无线的方式通过描述符来了解这些特征。节点层面的描述符包含节点描述符、电源描述符、复杂描述符和用户描述符。

节点描述符包含的内容很多，包括该节点设备类型(该节点是协调器、路由器或者终端设备)、制造商代码、是否选择用户描述符和复杂描述符、是否支持数据分片。

其他的一些描述符，如电源描述符，它是用来定义该节点支持的供电模式；用户描述符是以字符串的形式确定节点所在位置(如家里或者办公室)。这些描述符在 ZigBee 说明中进行选择。用户描述符能够在无线网络中设定，其他的描述符只能根据应用程序获得。

4 The Binding Manager performs the following:
- Establishes resource size for the Binding Table. The size of this resource is determined via a programmed application or via a configuration parameter defined during installation.
- Processes bind requests for adding or deleting entries from the APS binding table.
- Supports Bind and Unbind commands from external applications such as those that may be hosted on a PDA to support assisted binding. Bind and Unbind commands shall be supported via the ZigBee Device Profile.
- For the ZigBee Coordinator, supports the End Device Bind that permits binding on the basis of button presses or other manual means.

绑定管理执行下列任务：
- 为绑定表建立一个资源值。这个资源值是通过程序应用或通过一个在安装期间定义的配置参数确定的。
- 从 APS 绑定表增加或者减少实体处理绑定请求。
- 从外部应用支持绑定和解绑定命令，如那些是主机在一个 PDA 上来支持协助绑定。绑定和解绑定命令将通过 ZigBee 设备描述来支持。
- 对于 ZigBee 协调器，支持终端设备绑定，这绑定允许以按钮按压或其他手动菜单为基础的绑定。

第五节　网络层
Section 5　Network Layer

安全子域　security sub-field
断开网络　leaving a network
多播标志域　multicast flag sub-field
分布式地址分配机制　distributed address assignment mechanism
复位设备　resetting a device
功能实体　service entities
加入网络　joining a network
接收机同步　receiver synchronization
路径的发现和选择　route discovery and selection
路径保持维护　route maintenance
路径期满　route expiry
路由错误报告　route error reporting
路由发现　route discovery
路由协议　routing protocol
路由修复　route repair
命令帧　command frame
目的地址域　destination address field

数据帧的路由　routing of data frames
网络层管理实体　NWK layer management entity(NLME)
网络层数据实体　NWK layer data entity (NLDE)
网络发现　network discovery
网络信息库　network information base(NIB)
网络形成　network formation
网络最大深度　maximum depth in the network
信息库维护　information base maintenance
源地址域　source address field
源路由子域　source route sub-field
允许设备连接　allowing devices to join
帧格式　frame formats
帧控制域　frame control field
帧类型子域　frame type sub-field

1 The network layer is required to provide functionality to ensure correct operation of the IEEE 802.15.4-2003 MAC sub-layer and to provide a suitable service interface to the application layer. To interface with the application layer, the network layer conceptually includes two service entities that provide the necessary functionality. These service entities are the data service and the management service. The NWK layer data entity (NLDE) provides the data transmission service via its associated SAP, the NLDE-SAP, and the NWK layer management entity (NLME) provides the management service via its associated SAP, the NLME-SAP. The NLME utilizes the NLDE to achieve some of its management tasks and it also maintains a database of managed objects known as the network information base (NIB).

ZigBee 网络层的主要功能就是提供一些必要的函数,确保 IEEE 802.15.4-2003 的 MAC 子层正常工作,并且为应用层提供合适的服务接口。为了向应用层提供其接口,网络层提供了两个必要的功能服务实体,它们分别为数据服务实体和管理服务实体。网络层数据实体(NLDE)通过网络层数据服务实体服务接入点(NLDE-SAP)提供数据传输服务,网络层管理实体(NLME)通过网络层管理实体服务接入点(NLME-SAP)提供网络管理服务。网络层管理实体利用网络层数据实体完成一些网络的管理工作,并且,网络层管理实体完成对网络信息库(NIB)的维护和管理。

2 The default value for the NIB attribute nwkUseTreeAddrAlloc is TRUE, network addresses are assigned using a distributed addressing scheme that is designed to provide every potential parent with a finite sub-block of network addresses. These addresses are unique within a particular network and are given by a parent to its children. The ZigBee coordinator determines the maximum number of children any device, within its network, is allowed. Of these children, a maximum of nwkMaxRouters can be router-capable devices. The remaining devices shall be reserved for end devices. Every device has an associated depth which indicates the minimum number of hops a transmitted frame must travel, using only parent-child links, to reach the ZigBee coordinator. The ZigBee coordinator itself has a depth of 0, while its children have a depth of 1. Multi-hop networks have a maximum depth that is greater than 1. The ZigBee coordinator also determines the maximum depth of the network.

Given values for the maximum number of children a parent may have, nwkMaxChildren (Cm), the maximum depth in the network, nwkMaxDepth (Lm), and the maximum number of routers a parent may have as children, nwkMaxRouters (Rm), we may compute the function, Cskip(d), essentially the size of the address sub-block being distributed by each parent at that depth to its router-capable child devices for a given network depth, d, as follows:

$$Cskip(d) = \begin{cases} 1 + Cm(Lm - d - 1), & if\ Rm = 1 \\ \dfrac{1 + Cm - Rm - Cm \cdot Rm^{Lm-d-1}}{1 - Rm}, & otherwise \end{cases}$$

If a device has a Cskip(d) value of 0, then it shall not be capable of accepting children and shall be treated as a ZigBee end device for purposes of this discussion.

NIB 属性 nwkUseTreeAddrAlloc 的缺省值为真,采用分布式地址分配方案来分配网络地址,即该方案为每一个父设备分配一个有限的网络地址段。这些地址在一个特殊的网络中是唯一的,并且由它的父设备分配给它的子设备。ZigBee 协调器决定在其网络中允许连接的子设备的最大个数。对于这些子设备,参数 nwkMaxRouters 为路由器最大个数,而剩下的设备数为终端设备数。每一个设备具有一个连接深度,即连接深度表示仅仅采用父子关系的网络中,一个传送帧传送到 ZigBee 协调器所传递的最小跳数。ZigBee 协调器自身深度为 0,而它的子设备深度为 1。多跳网络的最大层深应大于 1。ZigBee 协调器决定网络的最大深度。

假定父设备拥有子设备数量的最大值为 nwkMaxChildren (Cm),网络的最大深度为 nwkMaxDepth (Lm),父设备将路由器最为它的子设备的最大数为 nwkMaxRouters (Rm),则可计算函数 Cskip (d),该函数为在给定网络深度 d 和路由器以及子设备个数的条件下,父设备所能分配子区段地址数为:

$$Cskip(d) = \begin{cases} 1 + Cm(Lm - d - 1), & Rm = 1 \\ \dfrac{1 + Cm - Rm - Cm \cdot Rm^{Lm-d-1}}{1 - Rm}, & 其他 \end{cases}$$

如果一个设备的 Cskip(d) 的值为 0,则它没有接收子设备的能力,并且将这样的设备看作为一个 ZigBee 网络的终端设备。

第六节 ZigBee 开发环境
Section 6 The ZigBee Development Environment

编译　compile
调试器　debugger
断点　breakpoint
仿真　simulating
仿真器　simulator
宏　macro

链接　link
片上系统　system-on-chip
设置选项　setting option
条件编译　conditional compilation
协议分析仪　protocol analyzer
运行　run

ZigBee is a small wireless networking protocol designed for IEEE 802.15.4 radios and 8-bit microcontrollers. Together, the radio, ZigBee stack, and microcontroller make what is called a platform. If the platform has been certified by the ZigBee Alliance as compliant (and not all ZigBee platforms are), then the platform is called a ZigBee Certified Platform, or ZCP.

Developing for ZigBee is basically the same for all platforms. A PC set of tools builds and compiles applications, which are then downloaded into target boards for debugging, usually through USB or Ethernet (although some platforms allow wireless download).

The PC tools required for ZigBee development usually include:

①An IDE for development, including compiling code into the appropriate form for the target microcontroller

②A ZigBee stack configuration or application "template" tool

③A debugger for downloading and stepping through source code lines on the target platform

④A protocol analyzer for debugging the network over-the-air

⑤Hardware for ZigBee development kits.

ZigBee 是一种专为 IEEE 802.15.4 无线标准和 8 位微控制器而设计的小型的无线网络协议。这种无线标准、ZigBee 协议栈和微控制器的集合称之为平台。如果该平台被 ZigBee 联盟认可(不是所有的 ZigBee 平台都被认证),则称该平台为 ZigBee 认证平台即 ZCP。

在所有的平台上进行 ZigBee 项目开发步骤基本上都是相同的。使用 PC 套件进行应用程序的编写和编译,接下来将代码下载到目标板上进行调试,下载过程通常使用 USB 或者以太网下载(有一些平台也支持无线下载)。

ZigBee 开发所需要的 PC 工具通常有:

①一种适用于目标微控制器代码编译的集成开发环境

②一个可配置的协议栈或者应用程序模板

③一个可以在目标平台上下载和单步调试代码的调试器

④一台可供无线网络调试使用的协议分析仪

⑤ZigBee 硬件开发套件

第十六章 误差理论与数据处理
Chapter 16　Error Theory and Data Processing

第一节　绪论
Section 1　Introduction

精度　accuracy　　　　　　　　　　修正值　correction
精密度　precision　　　　　　　　　引用误差　fiducial error
绝对误差　absolute error　　　　　　有效数字　significant figure
误差　error　　　　　　　　　　　　真值　true value
相对误差　relative error　　　　　　准确度　correctness

1 Error:

The measurement error is the deviation of the result of measurement from the true value of the measurable quantity, expressed in absolute or relative form.

If A is the true value of the measurable quantity and $\tilde{A}$ is the result of measurement, then the absolute error of measurement is $\zeta = \tilde{A} - A$. The error expressed in absolute form is called the absolute measurement error (error for short). The error expressed in relative form is called the relative measurement error.

误差：

观测值与被测量的真值之差称为误差,用绝对方式或相对方式表示。设 A 为真值,$\tilde{A}$ 为观测值,则绝对误差为 $\zeta = \tilde{A} - A$。测量误差可用绝对方式表示,称为绝对误差;绝对误差通常简称为误差。也可用相对方式表示,称为相对误差。

2 True value:

The true value of a measurable quantity is the value of the measured physical quantity. The true value of the measurable quantity is always unknown except for some particular cases. In practice, the value of the measurable quantity must be known with sufficient accuracy so that it can be used for this purpose instead of the true value of the quantity.

真值：

指在观测一个量时,该量本身所具有的真实大小。除了某些特定情况,真值一般是不知道的。实际测量中,常用满足规定精度的被测量的实际值来代替真值。

3 Relative error:

The relative error is the error expressed as fraction of the true value of the measurable quantity $\varepsilon = (\tilde{A} - A)/A$. Relative errors are nondenominational numbers normally given as percent.

相对误差：

绝对误差与被测量的真值之比值：$\varepsilon = (\tilde{A} - A)/A$。相对误差是无名数,通常以百分数(%)来表示。

4 Accuracy:

The accuracy of a measurement, corresponding quantitatively to the measurement error, reflects how close the result is to the true value of the measured quantity. For this reason, accuracy can be characterized quantitatively by a measurement error. A measurement is all the more accurate the smaller its error is, and vice versa.

精度：

反映测量结果与真值接近程度的量,通常称为精度。它与误差大小相对应,因此可用误差大小来表示精度的高低,误差小则精度高,误差大则精度低。

第二节　误差的基本性质与处理
Section 2　Basic Properties and Processing of Error

3σ 准则　3σ(treble mean square error) criterion
F 分布　F distribution
t 检验法　t test
χ^2 分布　χ^2 distribution
贝塞尔公式　Bessel formula
标准差　standard deviation
别捷尔斯法　Peters formula
不变的系统误差　constant systematic error
不同公式计算标准差比较法　standard deviation comparison of different formula
残余误差观察法　residual error inspection
残余误差校核法　residual error check
粗大误差　gross error
狄克松准则　Dixon criterion
反正弦分布　arcsine distribution
方差　variance
分布函数　distribution function
分布密度　distribution density
格罗布斯准则　Grubbs criterion
或然误差　probable error
极差法　range method
极限误差　limit error
计算数据比较法　calculation data comparison
加权算数平均值　weighted arithmetic mean
均匀分布　even distribution
罗曼诺夫斯基准则　Romanovskii criterion
平均误差　mean error
权　weight
三角形分布　triangle distribution
实验对比法　experimental comparison
算数平均值　arithmetic mean
随机误差　random error
随机误差特征　characteristic of random error
系统误差　systematic error
系统误差特征　characteristic of systematic error
线性变化的系统误差　linear systematic error
学生分布　student distribution
正态分布　normal distribution
秩和检验法　rank test
置信概率　confidence probability
置信水平　confidence level
置信系数　confidence coefficient
周期性变化的系统误差　periodical systematic error
最大误差法　maximum error method

1 Random error:
To define a random measurement error, imagine that some quantity is measured several times. If there are differences between the results of separate measurements, and these differences cannot be predicted individually and any regularities inherent to them manifested only in many results, then the error from this scatter of the results is called the random error.
Random errors are discovered by performing measurements of one and the same quantity repeatedly under the same conditions.

随机误差:
当对某量值进行多次测量时,得到一系列不同的测量值,每个测量值都含有误差,这些误差的出现没有确定的规律,不能单独预测,但就总体而言,却具有统计规律,这种误差称为随机误差。
随机误差可以通过在同一条件下,多次重复测量同一量值时予以发现。

2 Systematic error:
A measurement error is said to be systematic if it remains constant or changes in a regular fashion in repeated measurements of one and the same quantity.
Systematic error can be discovered experimentally either by comparing a given result with a measurement of the same quantity performed by a different method or by using a more accurate measuring instrument. However, systematic errors are normally estimated by theoretical analysis of the measurement conditions, based on the known properties of a measurement and of measuring instruments.
The observed and estimated systematic error is eliminated from measurements by introducing corrections. However, it is impossible to eliminate completely the systematic error in this manner.

Some part of the error will remain, and this residual error will be the systematic component of the measurement error.

系统误差：

在同一条件下，多次测量同一量值时，误差保持不变，或者在条件改变时，误差按一定的规律变化，这种误差称为系统误差。

系统误差可以通过以下几种方法发现：
(1)采用不同观测方法进行测量，将测量值互相对比以发现系统误差。
(2)用较高精度的仪器对同一量值进行测量，从而发现系统误差。
(3)理论分析法。根据对测量值和测量仪器特性的了解，通过分析观测条件来发现系统误差。

系统误差可以通过修正的方法予以消除，但用该方法不能将系统误差完全修正掉，总要残留少量系统误差成为测量误差的一个系统组成部分。

3 Cause and elimination of gross error:

When speaking about errors, we shall also distinguish gross or outlying errors and blunders. We shall call an error gross (outlying) if it significantly exceeds the error justified by the conditions of the measurements, the properties of the measuring instrument employed, the method of measurement, and the qualifications of the experimenter. Such measurements can arise, for example, as a result of sharp, brief change in the grid voltage (if the grid voltage in principle affects the measurements).

Outlying or gross errors in multiple measurements are discovered by statistical methods and are usually eliminated from analysis.

Blunders occur as a result of errors made by the experimenter. Examples are a slip of the pen when writing up the results of observations, an incorrect reading of the indications of an instrument, and so on. Blunders are discovered by nonstatistical methods, and they must always be eliminated from the analysis.

粗大误差产生的原因及消除方法：

粗差可以区分为粗大误差和错误两种。

在正常的测量环境、仪器、方法和人员素质下，测量误差应服从合理的分布。而当测量偏差明显超过应有的范围时，这种偏差称为粗大误差。例如，当电网电压（假设电网电压对该测量值有影响）意外急剧变化就会导致粗大误差的产生，这种粗差通常采用统计分析的方法予以发现并消除。

由于测量人员的主观原因（如记录时出现的笔误等）或对仪器进行错误读数可以记录为第二类粗差——错误，这种粗差的发现不能采用统计分析的方法，必须予以消除。

第三节　误差的合成与分配
Section 3　Synthesis and Distribution of Error

标准差的合成　combination of standard deviation
函数随机误差　random error of function
函数误差　function error
函数系统误差　systematic error of function
极限误差的合成　combination of limit error
偶然误差与系统误差的合成　combination of random error and systematic error
随机误差的合成　combination of random error
微小误差的取舍准则　criterion of accepting or rejecting small error

未定系统误差的合成　combination of undetermined systematic error
误差分配　error allocation
系统误差的合成　combination of systematic error
相关系数　correlation coefficient
已定系统误差的合成　combination of determined systematic error
最佳测量方案的确定　determination of optimal measure project

1 Function error and combination:

It's necessary to make indirect measurement when the direct measurement can not be made for particular measured quantity or when the accuracy of direct measurement is difficult to be ensured. The value of the indirect measurement quantity is calculated by directly measuring some other measurable quantities related to the the indirect measurement quantity when the functional relationship between the indirect measurement quantity and these measurable quantities is already known. Therefore, the value of the indirect measurement quantity is the function values of all the direct measurable quantities, and the error of the former is the function of all the errors of the latter. That is why the error of the former is called function error. The study of contents of the function error, in essence, is the study of the propagation of errors, of which, with determined relationship, the calculation is also called the error combination.

函数误差与合成：

由于被测量的特点，不能进行直接测量，或者直接测量难以保证测量精度时，必须采用间接测量。间接测量是通过直接测量与被测量之间有一定函数关系的其他量，按照已知的函数关系式计算出被测的量。因此间接测量的量是直接测量所得到的各个测量值的函数，而间接测量误差则是各个直接测得值误差的函数，故称这种误差为函数误差。研究函数误差的内容，实质上就是研究误差的传递问题，而对于这种具有确定关系的误差计算，也称之为误差合成。

2 Error allocation:

There are a number of errors in any measurement process. And the total error of the measurement results is the combined effect of various individual errors. The error allocation is adopted to ensure measurement accuracy, including measurement project selection according to the given allowance of the total error of the measurement results, reasonable error allocation and the individual error determination.

误差分配：

任何测量过程皆包含有多项误差，而测量结果的总误差则由各单项误差的综合影响所确定。误差分配是根据给定测量结果总误差的允差来选择测量方案，合理进行误差分配，确定各单项误差，以保证测量精度。

3 Criterion of accepting or rejecting small error:

An error is called the small error when its value is small to such a certain extent that it can be neglected in the calculation of the total error of the measurement results. For the random error and undetermined systematic error, practical calculation shows that an error can be rejected only if its value is less than or equal to 1/3 to 1/10 of the total standard deviation of the measurement results according to the criterion of accepting or rejecting small error.

微小误差的取舍准则：

当某误差数值小到一定程度后，计算测量结果总误差时可不予考虑，这种误差称为微小误差。实际计算表明，对于随机误差和未定系统误差，微小误差舍去准则是被舍去的误差必须小于或等于测量结果总标准差的 1/3～1/10。

4 Determination of optimal measurement project:

When the measurement results are related to various elements, what measures are taken to determine these elements so as to minimize the total error of the measurement results is the problem of determining the optimal measurement project. Generally the following aspects should be considered in determining the optimal measurement project:

(1)Select the optimal formula of function error.

(2)Make the error propagation coefficient equals to zero or a minimum.

最佳测量方案的确定：

当测量结果与多个测量因素有关时，采用什么方法确定各个因素，才能使测量结果的误差为最小，这就是最佳测量方案的确定问题。一般可从以下几方面来考虑：

(1)选择最佳函数误差公式。

(2)使误差传递系数等于零或为最小。

第四节 测量不确定度
Section 4 Measurement Uncertainty

A 类评定 estimation of type A uncertainty
B 类评定 estimation of type B uncertainty
标准不确定度 standard uncertainty
不确定度的报告 uncertainty report
测量不确定度的合成
combination of measurement uncertainty

测量不确定度 measurement uncertainty
合成标准不确定度
composite standard uncertainty
展伸不确定度 expanded uncertainty
自由度 degree of freedom

1 Measurement uncertainty:
Uncertainty of measurement is an interval within which a true value of a measurand lies with a given probability. Uncertainty is defined with its limits that are read out from a result of measurement in compliance with the mentioned probability.

测量不确定度:
是指被测量的真值以给定概率在某个量值范围内的一个估计,这种测量不确定度的定义表明,一个完整的测量结果应包含被测量值的估计与给定概率两部分。

2 Standard uncertainty estimation:
Standard uncertainty of measurement refers to the uncertainty characterized by standard. The components of measurement uncertainty are named as type A and type B uncertainties. The type A uncertainty is defined as a component of the measurement uncertainty that is estimated by statistical methods, whereas the type B uncertainty is estimated by nonstatistical methods.

标准不确定度评定:
用标准差表征的不确定度称为标准不确定度,测量不确定度的若干不确定度分量,均是标准不确定度分量,其评定分为 A 类评定和 B 类评定。A 类评定是用统计分析的方法进行评定,而 B 类评定不用统计分析法。

第五节 线性参数的最小二乘法处理
Section 5 The Least Square Method of Linear Parameters

测量方程 measurement equation
非线性参数 nonlinear parameter
精度估计 accuracy estimation
误差方程 error equation
线性参数 linear parameter

正规方程 normal equation
组合测量 compound measurement
最小二乘法原理
the principle of least square method

1 The least square method:
The least square method is a mathematical optimization technique. The optimal function can be found to match the measurement data best by minimizing the sum of the squared errors of the measured data and the estimated ones. The unknown data can be obtained easily by using least square method, and the sum of the squared errors of the measured data and the estimated ones is made minimum. The least square method is also used for fitting a curve to data points.

最小二乘法:
最小二乘法是一种数学优化技术。它通过最小化误差的平方来寻找数据的最佳函数匹配。利用最小二乘法可以简便地求得未知的数据,并使得这些求得的数据与实际数据之间误差的平方和为最小。最小二乘法还可用于曲线拟合。

2 The procedures of least square method of linear parameters:
First write out the error equations suitable to the specific issues. Then convert the error equations, based on the principle of least square method, into the normal equations by solving the extremum.

Finally, solve the normal equations to estimates the unknown parameters and their accuracy. For the nonlinear parameters, turn the nonlinear function into the linear form first by linearization, and then solve the linearized equations following the above-mentioned procedures of least square method of linear parameters.

线性参数的最小二乘法处理程序：
首先根据具体问题列出误差方程式；再按最小二乘法原理，利用求极值的方法将误差方程转换为正规方程；然后求解正规方程，得到待求的估计量；最后给出精度估计。对于非线性参数，一般采取线性化的方法，将非线性函数化为线性函数，然后按上述线性参数的最小二乘法处理程序去处理。

3 Compound measurement：
Compound measurement is to measure (generally in equal precision) the various combinations of the unknown parameters directly, and then the measurement data are treated in order to obtain the estimators of the unknown parameters and estimate their accuracy. Usually the measurement data of the combinations are processed by the least square method.

组合测量：
组合测量是直接测量待测参数的各种组合量（一般是等精度测量），然后对这些测量数据进行处理，从而求得待测参数的估计量，并给出其精度估计。通常组合测量数据是用最小二乘法进行处理。

第六节　回归分析
Section 6　Regression Analysis

多元线性回归　multiple linear regression
方差分析　analysis of variance
分组法　group method
函数关系　functional relationship
回归方程　regression equation
回归分析　regression analysis
回归直线　regression line
曲线回归方程　nonlinear regression equation
图解法　graphic method
显著性检验　significance test
线性递推回归　linear recursive regression
相关关系　correlation
相关指数　correlation index
一元非线性回归　one variable nonlinear regression
一元线性回归　one variable linear regression
一元线性回归方程　linear regression equation of one variable

1 Regression analysis：
Regression analysis is a mathematical technique for modeling and analyzing several variables, when the focus is on the relationship between a dependent variable and one or more independent variables.
回归分析：
回归分析就是应用数学方法，对若干变量进行分析处理，从而得出因变量和其他自变量之间关系的数学表达式。

2 Linear regression：
Linear regression is an approach to modeling the relationship between a scalar dependent variable y and one or more explanatory variables denoted X. The case of one explanatory variable is called simple regression or one variable linear regression. More than one explanatory variable is multiple linear regression. The model of the linear regression is normally given by the least square method.
线性回归：
线性回归是建立一个因变量和一个或多个自变量之间线性模型的方法，如果自变量只有一个，称为一元线性回归，如果是多个自变量，则称为多元线性回归。其求解一般采用最小二乘法来处理。

3 The simple methods for finding the the regression line：
Regression analysis, based on the least square method, is generally complex in calculation. For this reason, the following simple methods may be taken if the regression does not need to be so accurate or the linearity of the test data are good enough.
(1) The method of grouping (the average method)

Sort the explanatory variables ascending, and then divide the measurement equations into two (equal to the assumed number of unknowns) equal or nearly equal groups. Add the two groups respectively to get two equations related to the regression coefficients, and the regression coefficients and then the regression equation can be obtained by solving the set of equations.

(2) The graphic method (tight-rope method)

Draw a scatter graph on the coordinate paper with measurement data. Then draw a straight line equably through the point group if the point group appears to be a linear band. This line can be approximately regarded as the regression line, and the regression coefficients can then be obtained directly from the graph.

回归直线的简便求法：

回归分析是以最小二乘法为基础，计算一般比较复杂，在精度要求不太高或试验数据线性较好的情况下，可采取如下简便方法：

(1)分组法(平均值法)

将自变量数据由小到大排序，分成个数相等或近于相等的两组(分组数等于欲求的未知数个数)；分别将两组观测方程相加，得到关于两个回归系数的方程组；解此方程组即得回归方程。

(2)图解法(紧绳法)

把观测数据画出散点图于坐标纸上，假如画出的点群形成一直线带，就在点群中画一条直线，使其均匀穿过点群。这条直线可以近似地作为回归直线，回归系数可以直接由图中求得。

4 One variable nonlinear regression:

If the relationship between a dependent variable and an independent variable is not linear, but of some kind of curve, this regression is called the nonlinear regression. The procedures of one variable nonlinear regression are generally performed as follows:

(1) Determine the function type.

(2) Solve the unknown parameters of the correlation function. Linearize the regression curve by variable substitution to estimate the unknown parameters using linear regression above-mentioned, or expand the function into a polynomial so as to transform the solution of regression curve into that of polynomial regression.

一元非线性回归：

若两个变量之间不是线性关系，而是某种曲线关系，则属于一元非线性回归问题，一般可分两步进行：

(1)确定函数类型。

(2)求解相关函数中的未知参数。通常是通过变量代换把回归曲线转换成回归直线，进而按线性回归的方法求解；或者把回归曲线展成回归多项式，把解回归曲线问题转化为解多项式回归问题。

5 Linear recursive regression:

Linear recursive regression is suitable for dynamic measurement data processing. Firstly, the initial regression coefficients are estimated according to the initial measurement data. With the new data measured, the increments of the regression coefficients are calculated then, and the sum of the initial values of the regression coefficients and their increments return the new solution of the regression coefficients. When more new measurement data come, the similar calculation of increments starts once more. The calculation speed of the linear recursive regression coefficients is greatly improved because of less and non-repeated calculation of the increments.

线性递推回归：

该方法用于动态测量数据处理。首先根据初始的测量数据计算出回归系数初始值；新增加一组数后，计算出新增数据带来的回归系数增量，回归系数初始值加上其增量就是回归系数新的解；再增加新数据时，按此法类推计算。由于回归系数增量的计算工作量较少，而且无重复性计算，提高了计算速度。

第十七章 印刷电路板设计
Chapter 17　Design for the PCB

第一节　概述
Section 1　Introduction

AD6 网络升级　Altium web update
Altium 公司印刷电路板设计软件,版本号 6.x,简记:AD6　Altium Designer 6.x
CAM 编辑器　CAM editor
版本控制　version control
备份　backup
波形编辑器　wave editor
不具有电气意义的线　graphical line
导航　navigation
电子设计自动化　electric design automation (EDA)
动态透明效果　dynamic transparency
工程设计自动化　engineering design automation (EDA)
环境参数设置　preference
计算机辅助测试　computer-aided test(CAT)
计算机辅助设计　computer-aided design (CAD)
计算机辅助制图　computer aided drawing
计算机辅助制造　computer-aided manufacturing (CAM)
接受授权协议　accept the license agreement
描述系统　scripting system
目的文件夹　destination folder
嵌入式系统设计　embedded system
全屏显示　full screen
授权管理　licenses management
授权激活　active license
授权协议　license agreement
缩放精度　zoom precision
透明度　transparency
透明浮动窗口　transparent floating window
网络标签　net label
网络端口　port
文本编辑器　text editor
文件类型　file type
文件锁定　file locking
显示文件结构　show document structure
显示项目结构　show project structure
显示栅格　show grid
项目面板　project panel
新文档默认项　new document default
要显示的对象　object to display
页面标志　sheet symbol
页面端口　sheet entrie
页面接口　sheet connector
已安装的库　installed library
引脚　pin
隐藏栅格　hide grid
印刷电路板编辑器　PCB editor
印制板电路　printed circuit board (PCB)
用户信息　user information
重新打开上一次的工作空间　reopen last workspace

1 When you select All Programs ≫ Altium Designer 6 from the Windows Start menu to run Altium Designer, you are actually launching DXP.EXE. The DXP platform underlies Altium Designer, supporting each of the editors that you use to create your design.

当你从 Windows 开始菜单选择 All Programs ≫ Altium Designer 6 启动程序时,你实际上是启动了 DXP.EXE。DXP 平台位于 Altium Designer 之下,支持你所设计的每一个编辑器。

2 To display a document in AD6 editor, double-click on a document icon in the Projects panel. The document will be opened in the appropriate editor, e.g. Schematic Editor, PCB Editor or the Library Editors.

要在 AD6 编辑器中显示文档,在项目面板里双击文档的图标即可。文档将在相应的编辑器里被打开,例如:原理图编辑器、PCB 编辑器或库编辑器。

第二节　原理图设计
Section 2　Schematic Diagram Designing

半径　radius
保存设计空间　save design workspace
贝塞尔曲线　Bezier
比例因子　scaling factor
捕捉栅格　snap grid
参数　parameter
层次设计　hierarchical design
撤销/重做　undo/redo
导线布线　wiring
电路仿真　circuit simulation
电气栅格　electrical grid
电源地　power ground
端口信号流向　port direction
仿真　simulation
放元件　place part
公制单位　metric unit system
绘图应用　utility
混合信号仿真　mixed simulation
检查工具　inspector
矩形填充　rectangle filling
可视背景栅格　visible grid
可用的库　available library
连接点符号　junction
浏览库元件　browse library
逻辑电路　logic circuit
逻辑仿真　logic simulation
逻辑设计　logic design
逻辑图　logic diagram
模板　template
模块化　modularization
默认空白图纸尺寸　default blank sheet size
起始角度　start angle
器件封装　footprint
扫描填充　scan filling
扇形图　pie chart
设计规则检查　design rule checking(DRC)
设计数据库　design database
设计原点　design origin
设置元件时切断导线　component cut wire
时序仿真　timing simulation
实体设计　physical design
使用或取消电气栅格　toggle electrical grid
视图自动移动功能　auto pan option
输出文件路径　output path
鼠标轮设置　mouse wheel configuration
数字的　numeric
填充域　region filling
图形显示　graphics display
图纸连接端口的信号流向　sheet entry direction
网络标签　net label
网络表选项　net list option
文件输出选项　output option
系统字体　system font
显示或隐藏栅格　toggle visible grid
线网络　net
项目选项　project option
信号地　signal ground
修改过的文档　modified document
引脚的信号流向　pin direction
英文字母　alpha
英制单位　imperial unit system
优化连线　optimize wire
元件安置　component positioning
元件标号　designator
元件属性　component property
原点　origin
原理图　schematic diagram
原理图标准　schematic standard
原理图参数设置　schematic preference
栅格　grid
折线　polyline
正交拖动　drag orthogonal
终止角度　end angle
转换连线交叉点为连接点符号　convert cross junction
自底向上设计　bottom-up design
自顶向下设计　top-down design
自动放置电气连接点　auto junction
自动缩放　auto zoom
总线　bus
总线宽度　bus width
总线引入线　bus entry

1 The Schematic Editor opens when you open an existing schematic document or create a new one. This editor makes use of all the workspace features in the Altium Designer environment. This includes

multiple toolbars, resource editing, right-click menu, shortcut keys and Tool Tips.
当你打开一个已存在的原理图文件或新创建一个原理图文件时,原理图编辑器即被打开。该编辑器可以使用 AD6 环境下的所有工作空间性能。包括:多工具栏、源编辑、右键菜单、快捷键和工具小提示等。

2. Click and hold down the right mouse button and move the cursor to pan in a design document. The hand-shaped cursor indicates you are in panning mode. Release the right mouse button to stop panning.
点击鼠标右键并保持,同时移动光标就可在设计文档上平移窗口。手形光标指示你平移的方式。松开右键即可停止平移。

3. Use the Drawing Tools available on the Utilities toolbar to place the graphical objects. Turn the Utilities toolbar on and off by selecting View ≫ Toolbars ≫ Utilities.
在应用工具栏中使用绘图工具可以放置图形对象。通过菜单项 View ≫ Toolbars ≫ Utilities 可以打开或关闭应用工具栏。

4. Drawing toolbar functions can also be accessed through the Place ≫ Drawing Tools menu, except for Paste Array (Edit ≫ Smart Paste).
除了粘贴矩阵(Edit ≫ Smart Paste)之外,绘图工具栏通过 Place ≫ Drawing Tools menu 也能实现其功能。

第三节 原理图编辑
Section 3 Schematic Diagram Editing

编号索引控制   designator index control
查找相类似对象   find similar object
垂直等间距排列   distribute vertical
从组合中解除对象   break objects from union
当前文档中的所有对象   all on current document
对被选对象创建对象组合   create union from selected object
对齐到栅格   align to grid
对象组合   union
加标注   annotate
剪贴板   clipboard
解除被选的对象   deselect
排列与对齐   align
区域内对象的选择   inside area
区域外对象的选择   outside area

水平等间距排列   distribute horizontal
顺时针旋转被选对象   rotate selection clockwise
图形编辑   graphical editing
拖拽被选的对象   drag selection
网络表   net list
先由上至下、再按左至右   down then across
先由下至上、再按左至右   up then across
先由左至右、再按上至下   across then down
先由左至右、再按下至上   across then up
橡皮图章工具   rubber stamp
旋转被选的对象   rotate selection
移动被选的对象   move selection
以阵列形式粘贴   paste array
执行顺序   order of processing
智能粘贴   smart paste

1. To place a component, double-click on its name in the Libraries panel. To edit a component's properties before you place it, press the TAB key. The component properties dialog displays. To step through the fields in the dialog press TAB (down), or SHIFT+TAB (up).
要放置一个元件,在原理图库面板中双击该元件名称。在放置该元件之前要编辑元件属性,按下 TAB 键,显示出该元件属性对话框。要单步调试对话框区域,按 TAB 向下,或按 SHIFT+TAB 向上。

2. Custom Style Section allows you to define a custom sheet size and border. Use this option if you want a sheet size not covered in the Standard Style section.
自定义图纸类型允许你自定义图纸尺寸和边框。如果你所要图纸不在标准类型之列时可以使用这一选项。

3. The Snap Grid forces the mouse click location to the closest snap grid point. The Snap Grid is set and

can be turned on or off in the Document Options dialog. You can also cycle though three predefined grids by pressing the G shortcut key at any time.

捕捉栅格强制鼠标定位于离捕捉栅格点最近的位置。捕捉栅格在文档选项对话框中进行设置并可以打开或关闭。任意时刻通过按 G 快捷键你也可以在三种预定义栅格中循环打开。

第四节 原理图库文件编辑
Section 4 Schematic Library Editing

编译集成元件库　compile integrated library
电气类型　electrical type
集成元件库　integrated library
说明　description
引脚属性　pin property
元件引脚编辑器　component pin editor
原理图库　schematic library

1 The supplied components are stored within a set of integrated libraries. An integrated library includes the schematic symbols, plus it can also include all associated models, such as footprints, spice models, signal integrity models, and so on. Most of the supplied integrated libraries are manufacturer-specific.

提供的元件存储在集成库内。集成库包含原理图符号，也包括所有相关联的模块，例如封装、电路级仿真模块、信号整体性分析模块等。提供的大多数集成库都是由开发商特定的。

2 Integrated libraries are compiled from separate source schematic libraries, such as PCB footprint libraries, etc. The components in an integrated library cannot be edited, and to change a component the source library is edited and recompiled to produce an updated integrated library.

集成库是从分散的原理图库编译而来的，例如 PCB 封装库等。集成库中的元器件不能被编辑的，要改变元件需要在源库中编辑该元件并重新编译才能产生一个新的集成库。

第五节 PCB 设计
Section 5 Printed Circuit Board Designing

安全间距　clearance
板层摆放管理　layer stack manager
边界空白区域的宽度　keep out distance from board edge
边界线宽度　boundary track width
标题框和刻度栏　title block and scale
标注线宽度　dimension line width
薄膜开关　membrane switch
布局　placement
布线　routing
布线方案　routing strategy
布线完成率　layout efficiency
单面印制板　single-sided printed board
当前图层　current layer
导线面　conductor side
底层　bottom layer
底层丝印层　bottom overlay
电路板布线规则　PCB rule
电路板查看显示　board insight display
电源层　power plane
顶层　top layer

顶层丝印层　top overlay
多边形敷铜层　place polygon plane
多层印制板　multilayer printed board
多层印制电路板　multilayer printed circuit board
方焊盘　square pad
防焊膜　solder mask
光标位置　cursor location
过导孔　via
过孔　via hole
过孔类型选择　choose via style
焊接面　solder side
焊盘　pad
焊盘孔　pad hole
混合电　hybrid circuit
交互式布线　interactive routing
角度标注　angular dimension
镜像　mirroring
泪滴焊盘　teardrop pad
埋孔　buried via hole
盲孔　blind via (hole)
切除 PCB 板边角　corner cutoff

第十七章 印刷电路板设计

热键　hot key
设计规则检验　design rule check(DRC)
设计规则约束　design rule constraint
双面印制板　double-sided printed board
丝印层　silkscreen overlay
添加电源平面　add plane
添加信号层　add layer
通孔　thru-hole var

线性标注　linear dimension
选择 PCB 板轮廓　choose board profile
选择电路板的层数　choose broad layer
元件面　component side
圆焊盘　round pad
重布　rerouting
自动布线　auto route

1 The PCB Editor opens when you open or create a PCB document. It shares all the workspace features offered by the Altium Designer environment.
　当你打开或新建一个 PCB 文件时,PCB 编辑器即刻打开。它共享了 Altium Designer 环境所提供的所有工作空间性能。

2 A PCB is fabricated as a series of layers, including copper electrical, insulation, protective masking, text and graphic overlay layers.
　PCB 就是把一系列的板层组装在一起,包括电气铜层、绝缘层、保护层、字符层、图形丝印层等。

3 To start Interactive Routing, select the toolbar button or Place ≫ Interactive Routing (PT). Click where you wish to begin the first track and then use the track placement and start/end modes.
　要开始交互式布线,选择工具栏的交互式布线按钮或选菜单 Place ≫ Interactive Routing。点击你想开始第一段导线的位置,然后使用导线放置起始、终止模式。

4 Pads are mainly used as part of components but can be used as individual objects, such as testpoints or mounting holes.
　焊盘主要被用来作为元件部分,但也被用作为个别对象,例如测试点或安装孔。

5 Routing is the process of laying tracks and vias on the board to connect the components. Altium Designer makes the job of routing easy by providing a number of sophisticated manual routing tools as well as the new powerful Situs topological autorouter, which optimally routes the whole or part of a board at the touch of a button.
　布线就是在电路板上放置导线和过导孔来连接到元件的过程。AD6 提供了许多成熟的手工自动布线工具以及新的强有力的位置拓扑自动布线,所有这些,点一下按钮,就可以在全部或部分电路板实现最优化的布线,使得布线工作变得更容易。

第六节　PCB 设计规则设置
Section 6　PCB Design Rules

编辑规则属性　edit rule priority
布线拓扑类型　routing topology
布线优先次序　routing priority
测试点　testpoint
差分对布线　differential pairs routing
导出规则　export rule
导入规则　import rule

恢复布线　unroute
扇出式布线控制　fallout control
设计规则和约束编辑器　PCB Rules and Constraints Editor
推荐宽度　preferred width
信号完整性　signal integrity
约束　constraint

1 Schematic diagrams in Altium Designer are more than just simple drawings——they contain electrical connectivity information about the circuit. You can use this connectivity awareness to verify your design. When you compile a project, Altium Designer checks for errors according to the rules set up in the Error Reporting and Connection Matrix tabs and any violations generated will display in the Messages panel.
　AD6 原理图不仅仅是简单的绘制,还包含有关电路的电气连接信息。您可以使用此连接来验证您的设计。当您编译一个项目时,AD6 依照设计规则的设置来检查错误,形成错误报告和连接矩阵表，任何违反规则的都将被显示在信息面板中。

2 In Altium Designer, design rules are used to define the requirements of your design. These rules cover every aspect of the design——from routing widths, clearances, plane connection styles, routing via styles, and so on. Rules can be monitored as you work and you can also run a batch test at any time and produce a DRC report.

AD6 中设计规则被用来定义你的设计需求。这些规则涵盖了设计的各个方面，从布线宽度、容差、平面连接形式、布线过导孔形式等等。规则在你工作时能够被监视并且任何时刻能够运行批量测试并产生一个设计规则校验报告。

3 To effectively apply the design rules, the concepts of rule type, object set, query and priority need to be understood.

要充分利用设计规则，需要理解规则类型、对象集、查询和优先级等概念。

4 Design rules are checked by the Design Rule Checker (DRC) either online as you work, or as a batch process (with an optional report). The batch mode can be run at any time, and it is good design practice to run it as a final verification check when the board is completed.

不管是你在线工作还是批量处理(用一个可选报告)，设计规则都是由设计规则检查器来检验的。任何时刻都能运行批量模式，当电路板完成设计时运行它作为最后的检验是很好的设计惯例。

第七节　PCB 元件封装库编辑
Section 7　PCB Library Editing

X 方向间隔　X-spacing　　　　　　设置粘贴阵列　setup paste array
Y 方向间隔　Y-spacing　　　　　　双列直插封装　Dual In-Line Package(DIP)
电路板选项　broad option　　　　　线性阵列　linear array
封装　package　　　　　　　　　　元件　component
库选项　library option

1 The PCB Library Editor is used to create and modify PCB component footprints and manage PCB component libraries. The PCB Library Editor also includes a Component Wizard that you can guide through the creation of most common PCB component types.

PCB 库编辑器用于创建和修改 PCB 元件封装并管理 PCB 元件库。PCB 库编辑器也包括元件封装向导，通过向导你可以完成许多通用 PCB 元件类型的创建。

2 The PCB Library panel of the PCB Library Editor panel provides a number of features for working with PCB components.

PCB 库编辑器面板的 PCB 库面板提供随 PCB 元件工作的许多功能。

3 The PCB Library Editor includes a Component Wizard. This Wizard allows you to select from various package types, fill in appropriate information and it will then build the component footprint for you.

PCB 库编辑器包括元件封装向导，它允许你选择不同的封装类型、填充相应信息，然后它将为你建立元件封装库。

4 There will be situations where you need to create a footprint with pads that have an irregular shape. This can be done using any of the design objects available in the library editor.

将有一些情况，你需要用不规则焊盘形状创建一个封装。使用在库编辑器中可用的任何设计对象就可以做到。

第十八章 运动控制系统
Chapter 18 Motion Control System

第一节 闭环控制的直流调速系统
Section 1 DC Speed Regulation System of Closed-loop Control

PI调节器 proportional plus integral regulator
半桥整流电路 half bridge rectifier
比较电压 compare voltage
比例积分微分(PID)控制 proportional - integral- derivative control
比例控制 proportional control
闭环 closed loop
闭环放大倍数 closed-loop amplification factor
变压器漏抗 transformer leakage reactor
补偿控制 compensation control
测速发电机 speed indicating generator
触发装置 trigger equipment
电磁时间常数 electromagnetic time constant
电磁转矩 electromagnetic torque
电力半导体器件 power semiconductor device
电力变流器 power converter
电力公害 power public hazard
电力拖动 electric drive
电流反馈 current feedback
电流反馈系数 current feedback factor
电流负反馈 current reversed feedback
电流截止负反馈 current cut-off reversed feedback
电流正反馈 current forward feedback
电枢电压 armature voltage
电枢电阻 armature resistance
电网电压 electric network voltage
电压反馈 voltage feedback
电压反馈系数 voltage feedback factor
电压负反馈 voltage reversed feedback
电压负反馈控制系统 voltage reversed feedback control system
调速范围 speed regulation range
调速系统 speed regulation system
调速指标 speed regulation index
调压调速系统 voltage and speed regulation system
动态数学模型 dynamic mathematical model
额定电流 rating current
额定电压 rating voltage
额定负载 rating load
额定速降 rating speed drop
反馈检测 feedback detection
反馈检测精度 feedback detecting accuracy
反馈系数 feedback factor
非稳态系统 astable system
幅值控制 amplitude control
负载电流 load current
负载转矩 load torque
感应电动机 induction motor
给定电压 given voltage
惯性系数 inertia constant
过补偿 over compensation
过电流 over-current
过电压 over-voltage
回馈制动 feedback brake
霍耳电流变换器 Hall current converter
机电时间常数 electromechanical time constant
机械特性 mechanical characteristic
积分调节器 integral regulator
积分控制 integral control
鉴幅器 amplitude discriminator
交流 alternating current (AC)
交流电机 AC motor
交流电气传动 AC electric drive
交流互感器 AC transformer
截止电流特性 cut-off current characteristic
截止频率 cut-off frequency
晶闸管 thyristor
静差率 slip rate
开环 open loop
开环放大倍数 open-loop amplification factor
抗扰性能 anti-interfere performance
空载电流 idle current
理想空载 ideal no-load
理想空载转速 ideal free-wheeling rotation
励磁磁通 excitation flux
临界放大倍数 critical amplification factor
脉冲宽度调制 pulse width modulation
脉冲频率调制 pulse frequency modulation

门极触发信号　gate trigger signal
内部电压降　internal drop
逆变变压器　inverter transformer
逆变电源　inverter power supply
逆导晶闸管　reverse conducting thyristor
偏差电压　bias voltage
平波电抗器　flat wave reactor
欠补偿　under compensation
强迫关断电路　forced turn-off circuit
全补偿　overall compensation
全桥整流电路　full bridge rectifier
瞬时电压　transient voltage
四象限运行　four-quadrant running
速度反馈　speed feedback
通用电动机　all purpose motor
同步变压器　synchronous transformer
微分控制　differential control
稳态跟随误差　steady-state track error
稳态结构图　steady-state structure diagram
稳态抗扰误差　steady-state anti-perturbation error
稳态速降　steady-state speed drop
稳态误差　steady-state error
系统框图　system block diagram
系统设计　system design
系统稳定性　system stability

限幅器　amplitude limiter
限流保护　limited current protection
限流装置　current limiting device
相位控制　phase control
谐波分析　harmonic analysis
续流二极管　freewheeling diode
旋转变压器　magslep
有静差调速系统　speed regulation system of steady-state error
整流变压器　rectifier transformer
整流电路　rectifier circuit
整流滤波器　rectifier filter
整流平均值　rectified mean value
整流器　rectifier
整流纹波系数　rectified ripple factor
正向压降　forward voltage drop
直流　direct current(DC)
直流电机　DC motor
直流电气传动　DC electric drive
直流互感器　DC transformer
直流励磁发电机　DC excitation generator
直流斩波器　DC chopper
滞后时间常数　delay time constant
转动惯量　freewheel inertia
自动控制系统　automatic control system

1 Open-loop system: A system is not self-correcting when variations in input and process conditions occur spontaneously. This system requires human monitoring.
开环系统：当输入或过程条件发生变化时，系统不能自动调节。该系统需要人为控制。

2 PID control is the abbreviation of proportional plus integral plus derivative control.
比例积分微分控制是比例加积分加微分控制的缩写形式。

3 Unstable system: A system in which oscillations build up is said to be unstable and much of the design work in control engineering is associated with producing a stable system.
不稳定系统：存在振荡状态的系统称为不稳定系统，控制工程的许多设计工作都与创建稳定系统有关。

第二节　双环控制的直流调速系统
Section 2　DC Speed Regulation System of Dual-loop Control

饱和电压　saturation voltage
饱和非线性控制　saturation non-linear control
闭环传递函数　closed-loop transfer function
闭环极点　closed-loop pole
闭环零点　closed-loop zero
参数配合　parameter match
超调量　overshoot
典型系统　typical system
电磁转矩　magnetic torque

电动势调节器　electromotive force regulator
电流调节器　current regulator
电流互感器　current transformer
电流环　current loop
电容滤波　capacitor filtering
调节时间　settling time
动态速降　dynamic speed drop
动态性能　dynamic performance
堵转电流　locked-rotor current

第十八章　运动控制系统

二阶最佳	second-order optimum	输出限幅	output amplitude limit
反馈滤波	feedback filter	衰减系数	decaying coefficient
峰值时间	peak time	双闭环调速系统	double closed-loop speed regulation
给定滤波	given filter	双稳态	bistability state
跟随性能	follow performance	速度环	speed loop
惯性	inertia	特征值	characteristic value
恢复时间	recovery time	微分负反馈	differential reversed feedback
角频率	circular frequency	稳定	stability
阶跃输入	step input	稳定条件	steady-state condition
截止频率	blocking frequency	稳态响应	steady-state response
开环传递函数	open-loop transfer function	限幅电路	clipping circuit
抗扰性能	anti-interference performance	振荡	oscillation
控制对象	control object	转速超调	rotation overshoot
励磁电流调节器	magnetic current regulator	转速调节器	rotation regulator
启动过程	start procedure	转速负反馈	speed reversed feedback
弱磁控制	weak magnet control	准时间最优控制	quasi-time optimum control
三阶最佳	third-order optimum	准稳态	quasi-stable state
上升时间	rise time		

1 Response time (tr): The time the output variable of the control system takes from the application of the input step change to 90% of the steady-state value.

响应时间(tr)：控制系统在阶跃输入作用下，输出变量上升到稳态值的90%所用的时间。

2 Settling time (ts): The time the output variable takes to settle within a small percentage (2%-5%) of the steady-state value.

调节时间(ts)：输出变量达到稳态值的较小百分比(2%～5%)所用的时间。

3 Overshoot: The overshoot in transient response is expressed as the percentage of the output variable to the steady-state value.

超调量：瞬时响应的超调量为输出变化量与稳态值的百分比。

4 Steady-state error is the error between the reference input and the actual steady-state value of the output variable.

稳态误差是输出变量的实际稳态值与参考输入之间的误差。

第三节　可逆调速系统
Section 3　Reversible Speed Regulation

动态环流	dynamic circular current	空气开关	air-break switch
反并联线路	anti-parallel circuit	连锁保护	interlocking protection
反组触发装置	reversal trigger equipment	零电流检测器	zero current detector
封锁延时	lock delay	逻辑控制	logic control
环流	circular current	逻辑控制器	logic controller
环流电抗器	circular current reactor	脉冲封锁	pulse lock
击穿电流	breakdown current	弱磁控制	field-weakening control
击穿时间	breakdown time	正组触发装置	forward trigger equipment
静态环流	steady-state circular current	制动	brake
开放延时	open delay	制动转矩	brake torque
可逆线路	reversible circuit	转矩极性鉴别器	torque polar identifier
空气断路器	air circuit breaker		

1 Four-quadrant operation:
The four-quadrant operation refers to the operation of the DC motor with regenerative braking in both forward and reverse directions of rotation. In the first and third quadrants, the energy flows from the DC source to the load. The second and the fourth quadrant operations indicate regenerative braking where the kinetic energy of the motor is returned to the DC source.

四象限运行:
四象限运行指的是可以正反向旋转的直流电动机的运行,该电机可以实现再生制动。在一、三象限,电能从直流电源流向负载;在二、四象限,电机工作在再生制动状态,电机的动能被回馈到直流电源。

2 Regenerative Braking:
The DC motor operates in this mode when, drivng the forward motion of the motor, the polarity of the DC supply across the armature is reversed for stopping or reversing the direction of rotation. The motor acts as the generator. The kinetic energy of the motor is converted into electrical energy and returned to the source. The armature and field currents are opposite in polarity.

再生制动:
在电动机正向运行时,直流电动机制动时将电枢电源的极性反向或者改变旋转方向。电动机作为发电机工作,电机动能转变成电能然后回馈到电源,此时电枢和励磁电流极性相反。

3 Dynamic Braking:
The dynamic braking refers to removing the DC source and connecting a resistor R across the armature winding for stopping the motor. The kinetic energy of the motor is dissipated in R. The armature and field currents are opposite in polarity.

动态制动:
动态制动是指为了使电动机停止运转,在电机绕组两端并接电阻 R,而后断掉直流电源。电动机的动能由电阻 R 吸收,此时,电枢和励磁电流极性相反。

第四节 脉宽调制的直流调速系统
Section 4 DC Speed Regulation System of Pulse Width Modulation

安全工作区 safe operating area(SOA)	间接直流变换电路 indirect DC-DC converter
初始状态 initial condition	开关关断特性 switching turn-off characteristic
电力晶体管 giant transistor	开关瞬态过程 switching transient
电力系统分析 power system analysis	开关损耗 switching loss
电流连续模式 continuous conduction mode	开关特性 switching characteristic
断态(阻断状态) off-state	开关通态特性 switching turn-on characteristic
多重化 multiplex	可逆 PWM 变换器 reversible PWM converter
多重逆变电路 multiplex inverter	门极可关断晶闸管 gate turn-off thyristor
反向击穿 reverse breakdown	通态 on-state
供电电压 power-supply voltage	载波 carrier wave
缓冲电路 snubber circuit	斩波控制 chopper control
换路开关 circuit-changing switch	

1 Pulse width modulation (PWM), is a commonly used technique for controlling power to inertial electrical devices, made practical by modern electronic power switches. The average value of voltage (and current) fed to the load is controlled by turning the switch between supply and load on and off at a fast pace. The longer the switch is on compared to the off periods, the higher the power supplied to the load is.

脉冲宽度调制(PWM)是控制惯性电动装置电源的常用技术,通常由现代电力电子开关组成。提供给负载的电压(和电流)平均值由电源和负载之间的开关通断控制。开通时间与关断周期之比越长,供给负载的电源电压就越高。

2 The PWM switching frequency has to be much faster than what would affect the load, which is the device that uses the power. Typically switchings have to be done several times a minute in an electric stove, 120Hz in a lamp dimmer, from few kilohertz (kHz) to tens of kHz for a motor drive and well into the tens or hundreds of kHz in audio amplifiers and computer power supplies. The term duty cycle describes the proportion of "on" time to the regular interval or "period" of time; a low duty cycle corresponds to low power, because the power is off for most of the time. Duty cycle is expressed in percent, 100% being fully on.

PWM 开关频率必须高于使用该电源的负载响应。典型开关在电炉内是 1 分钟几次,调光灯 120 赫兹,电机驱动是几到几十千赫,音频放大器和计算机电源要几十到几百千赫。占空比是"开通"时间与有规律的间隔或周期时间的比值,占空比低输出功率就低,因为关断时间长。占空比用百分数来表示,全通是 100%。

3 The main advantage of PWM is that power loss in the switching devices is very low. When a switch is off there is practically no current, and when it is on, there is almost no voltage drop across the switch. Power loss, being the product of voltage and current, is thus in both cases close to zero. PWM also works well with digital controls, which, because of their on/off nature, can easily set the needed duty cycle.

PWM 控制技术的主要优点是开关装置的功耗低,当开关关断时电流几乎为零,而开通时开关两端几乎没有电压降。所以电流与电压乘积的功耗在开和关时基本为零。因为开/关的性质,PWM 很容易进行数字控制,也可以轻松地设置所需的占空比。

第五节　交流调压调速系统
Section 5　AC Speed Regulation System of Voltage Regulation

变频电动机	adjustable frequency motor	可变电容器	adjustable condenser
变频调速	variable frequency speed regulation	可调变压器	adjustable transformer
调速电机	adjustable-speed motor	可调电抗器	adjustable inductor
动作电流	actuating current	可调电阻器	adjustable resistor
换能器	transducer	可调恒速电动机	adjustable constant speed motor
交流电力电子开关	AC power electronic switch	全压启动	across-the-line starting
交流电力控制	AC power control	有功分量	active component
交流调功电路	AC power controller	有效负载	active load
交流调速	AC speed regulation	直接启动电动机	across-the-line motor
交流调压电路	AC voltage controller	转差功率	slip power

1 Traditionally, AC-AC conversion using semiconductor switches is done in two different ways: (1) in two stages (AC-DC and then DC-AC) as in DC link converters or (2) in one stage (AC-AC). Cycloconverters are used in high power applications driving induction and synchronous motors. They are usually phase-controlled and they traditionally use thyristors due to their ease of phase commutation. There are other newer forms of cycloconversion such as AC-AC matrix converters and high frequency AC-AC converters and these use self-controlled switches. These converters, however, are not popular yet.

一般情况下,使用半导体开关的交流-交流变频器有两种形式:一种是交流-直流-交流形式,中间含有直流变换器,另一种是交流-交流变换。周波变换器用于大功率驱动感应电机和同步电机的驱动。通常采用相控方式,由于换向简单一般采用晶闸管作为开关器件。其他新型周波变换器例如交流矩阵变换器、高频交流变换器采用全控器件,但这些变换器还没有普及。

2 Cycloconverter:

A cycloconverter is used in high power (>100 kWatts) applications. The thyristor is the device of choice. Natural commutation is the preferred method of switching the thyristors. Output frequency is lower than the source frequency. Its typical range is 0.2Hz to 15Hz. The amplitude and the frequency are both controllable.

周波变流器：
周波变流器应用于大功率(大于100千瓦)场合。装置的元件采用晶闸管。晶闸管的开关方式采用自然换向的方法。输出频率低于电源频率，典型范围为0.2赫兹到15赫兹。输出的幅值和频率均可控。

3 The three-phase cycloconverters are mainly used in AC machine drive systems running three-phase synchronous and induction machines. They are more advantageous when used with a synchronous machine due to their output power factor characteristics. A cycloconverter can supply lagging, leading, or unity power factor loads while its input is always lagging. A synchronous machine can draw any power factor current from the converter. This characteristic operation matches the cycloconverter to the synchronous machine. On the other hand, induction machines can only draw lagging current, so the cycloconverter does not have an edge compared to the other converters in this aspect for running an induction machine. However, cycloconverters are used in Scherbius drives for speed control purposes driving wound rotor induction motors.

三相周波变流器主要用于三相同步和感应电机的交流电机驱动系统。用在同步机上的优点是其输出功率因数高，周波变流器可以给滞后、超前或满功率因数负载供电，但其输入总是滞后的。这种变换器用于同步机可以牵引任何功率因数的负载电流，这一特点使得周波变换器用于同步电动机。另一方面，感应电机只能牵引滞后的负载电流，所以与其他变换器相比，在感应电机控制方面周波变换器并不具有优势。然而，在谢尔比斯驱动系统中周波变换器被用于控制感应电机转子绕组以达到调速的目的。

第六节　异步电动机变压变频调速系统
——转差功率不变型的调速系统

Section 6　Variable Voltage Variable Frequency Speed Regulation System for Asynchronism Motor
——Speed Regulation System of Slip-power Constant

SPWM 逆变器　sinusoidal PWM inverter
采样周期　sampling period
超前角　advanced angle
电流源逆变器　current source inverter
电压源逆变器　voltage source inverter
调制波　modulating wave
方波　square wave
给定积分器　given accumulator
规则采样法　regular sampling
函数发生器　function generator
交-交变频器　AC-AC frequency converter
交流并励电动机　AC shunt motor
交流潮流　AC power flow
交流串励电动机　AC series motor
交流分量　AC distribution
交流换向器电动机　AC commutator motor

交—直—交变频器　AC-DC-AC frequency converter
交—直流变流器　AC-DC converter
锯齿波调制　serrasoid modulation
绝对值变换器　absolute value converter
控制模式　control pattern
脉冲幅度调制　pulse amplitude modulation
脉冲平均电路　pulse-averaging circuit
脉冲限幅　pulse clipping
脉冲载波　pulse carrier
脉动因数　pulsation
三角波　triangular wave
压控振荡器　voltage controlled oscillator
载波　carrier wave
自然采样法　natural sampling
自适应控制　adaptive control

1 Voltage-to-frequency ratio (V/F):
When variable-frequency speed control of an AC motor is implemented, the motor supply voltage

must be adjusted to maintain a constant ratio——the V/F ratio.
压频比(V/F)：
当交流电动机采用变频调速时，电机的供电电压必须调整以维持一个恒定的比值——电压/频率的比。

2 Constant-torque operation：
In the region of speed below the rated value, the solid curves show the torque-speed characteristics at the low values of slip frequency. The source voltage is varied in proportion to the source frequency for holding the air-gap flux constant. The motor can deliver a constant torque at the rated value by drawing the rated current.
恒转矩工作：
在电动机机械特性固有特性中，额定转速以下区域，转差频率也低。所以为了保证电机气隙磁通恒定，电源电压与电源频率需要同比例变化，这样电动机产生一个和拖动电流相同比率的固定转矩。

3 Constant-power operation：
The motor speed can be increased beyond the rated value by increasing the frequency of the AC source connected to the stator winding. In this region the source voltage is held constant at the rated value. The ratio V/F decreases, consequently the air-gap flux decreases. The rotor current is constant, hence, the power loss in the rotor is also constant.
恒功率工作：
电机速度在定子交流电源频率提高时可以超出其额定转速，在这个区域电源电压保持恒定，压频比下降，相应气隙磁通也降低。此时，转子电流恒定，所以转子功率损耗也恒定。

第七节　绕线转子异步电动机串级调速系统
——转差功率回馈型的调速系统
Section 7　Cascade Speed Regulation System for Winding Motor
——Speed Regulation System of Slip-power Recovery

串级变频器　concatenated frequency converter
串级电机　concatenated motor
串级调速　cascade speed regulation
串级双馈电机　cascade doubly fed motor
附加电动势　additional electromotive force
功率变换单元　power converter unit
间接起动　indirect starting
静电感应晶体管　static induction transistor
静电感应闸管　static induction thyristor
静止无功补偿器　static var compensator
绕线电机　winding motor
矢量控制　vector control
鼠笼式电动机　squirrel-cage motor
双轴原型电机　two-axis primitive motor
提高功率因数　power factor improvement
同步电动机　synchronous motor
同步感应电动机　synchronous induction motor
直接起动　direct starting
直接转矩控制　direct torque control
转差功率　slip-power
转差功率回馈型调速系统　speed regulation system of slip-power recovery
转子电势　rotor electromotive
最大转矩　pull-out torque
最小转矩　pull-up torque

1 Slip is the difference between the synchronous speed and the rotor speed of an induction motor.
转差是感应电动机同步速度与转子速度的差值。

2 Line-frequency variable-voltage drive：The line-frequency variable-voltage drive is used for induction motors that have large rotor resistance and hence pull-out torque at large slip values. The drive permits speed control over a wide range.
恒压频比调速：恒压频比调速通常用于具有大转子阻抗的感应电动机，该电机在转差大时转矩也大。这种调速使转速可以在很宽的范围内调整。

[3] The double closed-loop cascade speed control system by chopper is designed to suit the demand of electric driving motor system for piercing mill. By introducing the basic principle of the speed control system using cascade control method with IGBT chopper, its mechanical characteristic is hard to be suitable for the periodically changed short time working load of the piercer. Furthermore, double closed-loop control system is designed for earning better static and dynamic performance in roll milling period. This method is worthy to be spread in engineering application.

为满足钻孔机的电机驱动要求,双闭环串级调速被设计成斩波器。通过引进由IGBT斩波器构成的串级调速系统的基本原理,其机械特性很难适合周期性改变的钻孔机短时工作负载。此外,为了获得辊磨机更好的动静态特性,设计了双闭环控制系统,这种方法在工程应用中值得推广。

第十九章 智能仪器设计
Chapter 19 Design of Intelligent Instruments

第一节 概述
Section 1 Introduction

倍增器	multiplier	门阵列	gate array
标准单元	standard cell	通信技术	communication technology
测量技术	measuring technology	网络化传感器	networked sensor
初级智能	primary intelligent	现代仪器	modern instrument
传感器	sensor/ transducer	信息技术	information technology
传统仪器	traditional instrument	仪器	instrument
聪敏	smart	智能传感器	smart sensor
高级智能	high-level intelligent	智能仪器	intelligent instrument
计算机技术	computer technology	自主测量仪器	autonomous measurement machine
聋哑传感器	dumb sensor		

1 "This magnitude places the earthquake as the fourth largest in the world since 1900 and the largest in Japan since modern instrumental recordings began 130 years ago", according to a USGS statement.

美国地质勘探局发布的声明说:"该强度使这次地震成为1900年以来全世界第四大地震,也是130年前开始用现代仪器记录地震以来在日本发生的最大地震。"

2 Intelligent instrument is the combination of the electronic measuring instrument and computer technology.

智能仪器是电子测量仪器和计算机技术紧密结合的产物。

3 The sensors are transforming the physical world into a computing platform. Modern sensors not only respond to physical signals to produce data, they also embed computing and communication capabilities. They are thus able to store, process locally and transfer the data they produce over a long distance.

传感器是把模拟的物理量转换为可测的数值量。随着现代技术的发展,传感器不仅可以转换模拟信号,而且还具有计算和通信的功能。因此,现代传感器可以在本地储存、处理数据,并能够远距离传送采集到的数据。

4 With ASIC, personal equipment, and other related technical development, intelligent instruments will be more widely used.

随着专用集成电路、个人仪器等相关技术的发展,智能仪器将会得到更加广泛的应用。

第二节 数据采集技术
Section 2 Data Acquisition Technology

保持	hold	调制技术	modulation technique
比较型模数转换器	comparative ADC	分辨率	resolution ratio
采集	acquisition	分布式数据采集	distributed data acquisition
采样	sampling	高通	high-pass
采样保持器	sampling holder	隔离放大器	isolation amplifier
采样抽取技术	sample extraction technology	光纤传感器	fiber optical sensor
采样误差	sampling error	过采样技术	over sample technique
程控增益放大器	gain-programmed amplifier	集成传感器	integrated transducer
低通	low-pass	集中式数据采集	centralized data acquisition

精度	precision	双积分型模数转换器	double integral ADC
绝对精度	absolute precision	系统误差	system error
量化	quantization	线性度误差	linearity error
滤波	filtering	相对精度	relative precision
偏移误差	offset error	信号调理	signal conditioning
前置放大	preamplification	仪用放大器	instrument amplifier
数据采集	data acquisition	增益误差	gain error
数据采集系统	data acquisition system	逐次逼近寄存器	successive approximation register
数字传感器	digital sensor		
数字滤波	digital filtering	转换时间	conversion time

1 Data acquisition system is one of the kernel components of digital signal processing.
数据采集系统是数字信号处理的核心部件之一。

2 Integrated temperature sensors are made by using silicon semiconductor integrated technologies, so also called silicon sensors or the single-chip integrated sensors.
集成温度传感器是采用硅半导体集成工艺而制成的,亦称硅传感器或单片集成传感器。

3 The measure module has bigger signal-to-noise ratio because adopting instrumentation amplifier, programmable gain amplifier and accumulate measurement.
信号检测模块采用仪表放大、程控增益放大、积累检测方式等措施有效地提高了测量系统的信噪比。

4 An appropriate sampling frequency is the key of the data acquisition system working reliably.
选择一个合适的采样频率是保证数据采集系统可靠工作的关键。

5 Recently years, with DSP techniques further developing, it has made successes to introduce the oversampling and decimation techniques into data acquisition system.
由于近来数字信号处理器(DSP)技术的进步,一种将过采样及抽取的概念引入数据采集系统的设计已取得成功。

第三节 人机对话与数据通信
Section 3 Human Computer Interaction and Data Communication

LCD 显示器	LCD display	即插即用	plug-and-play
编码键盘	encoded keyboard	键盘	keyboard
表面声波式触摸屏 surface acoustic type touch screen		静态驱动方式	static driving way
		控制传输	control transfer
触摸屏	touch screen	批量传输	bulk transfer
触摸屏技术	touch screen technology	热插拔	hot attach & detach
串行总线	serial bus	人机对话	human computer interaction
电平转换	level shift	设备	equipment
电容式触摸屏	capacitive touch screen	数据通信	data communication
电阻式触摸屏	resistive touch screen	通用串行总线	universal serial bus
调制解调	modulation-demodulation	无线数据传输	wireless data transmission
迭加驱动方式	superimposition driving way	线反转	line-reverse
非编码键盘	non-coding keyboard	行扫描	row-scanning
根集线器	root hub	中断传输	interrupt transfer
光点矩	opti-matrix	主机	host
红外线式触摸屏	infrared type touch screen	主控制器	master controller

1 In some defined working range, the man-computer dialogue interface can be adjusted on the computer screen, or can set all kinds of technical parameters of the machine.
人机对话界面,在界定的使用范围内,可在计算机屏幕上调整,或者设定机器的各种技术参数。

第十九章 智能仪器设计

2 Adopting large lattice LCD screen and intelligentized controlling system, a good English and Chinese language man-computer dialogue interface can be achieved.
采用大屏点阵液晶显示及智能化控制系统能提供良好的中英文人机对话界面。

3 Touch screens make computing feasible in new places, especially public ones, by doing away with keyboards, which can get gummed up with grime or spilled drinks.
触摸屏使计算在新的场所更简单易用，尤其是公共场所，由于没有键盘而不必担心染上污垢和碰到打翻的饮料。

4 USB protocol has four transfer types: Control Transfer, Bulk Transfer, Interrupt Transfer, and Isochronous Transfer.
USB 协议具有四种传输方式，即控制传输、批量传输、终端传输和等时传输。

5 In communication systems, there are mainly two methods for controlling transmission errors, one of which is forward-error-correction (FEC) and the other is automatic-repeat-request (ARQ).
数据传输系统中有两种技术被用来控制传输错误：即前向差错控制技术(FEC)和自动要求重传技术(ARQ)。

第四节 智能仪器的基本数据处理算法
Section 4 Basic Data Processing Algorithms of the Intelligent Instruments

标度变换	scale conversion	曲线拟合	curve fitting
粗大误差	parasitic error	去极值	removing extreme value
代数插值	algebraic interpolation	数据处理	data processing
非线性校正	nonlinear correction	算法	algorithm
分段曲线拟合	piecewise curve fitting	算术平均值	arithmetic mean
分段线性插值	piecewise linear interpolation	随机干扰	random disturbance
高频噪声	high-frequency noise	随机误差	random error
滑动平均	moving average	温度误差	temperature error
加权	weighting	系统误差	system error
绝对偏差	absolute deviation	相关分析	correlation analysis
零位误差	zero error	校正函数	correcting function
脉冲干扰	impulse interference	增益误差	gain error
频谱估计	spectrum estimation		

1 The technologies including nonlinear compensation, scale conversion, algorithms of signal preprocessing, FFT spectrum analyzing are emphasized.
着重阐述了数据的非线性补偿技术、标度变换技术、信号预处理的各种算法、FFT 谱分析技术。

2 Frequency estimation and spectrum analysis are the foundation of the analysis of complex communication signals in electronic reconnaissance system.
频率估计和频谱分析是在电子侦察系统中进一步分析复杂通信信号的基础。

3 To restrain random interference and pulse interference and improve measurement accuracy and stability, the digital filtering method which the average data crossing the extremum out is adopted.
数据处理中，采用了去极值平均滤波的数字滤波方法，有效地抑制了随机干扰及脉冲干扰，增加了系统的稳定性，提高了测量精度。

4 In order to insure the veracity of the data from software as much as possible, it introduces the median filter and conic approach technique in dealing with measuring data.
在测量数据的处理中，采用了中值滤波与二次曲线拟合技术，从软件上尽可能保证测量值的准确性。

第五节 软件设计
Section 5　Design of Software

编码　encoding
反推工程　reverse engineering
工程软件　engineering software
工具软件　tools software
过程　procedure
过程开发模型　process development model
后台调试模式　background debugging mode
集成测试　integrated testing
计算机辅助软件工程
computer-aided software engineering
继承性　inheritance
监控程序　monitor program
交叉开发　cross developing
开发　development
可重用软件　reusable software
螺旋模型　spiral model
面向对象　object-oriented
模块化设计　modular design
目标机　target
瀑布式模型　waterfall model
嵌入式软件　embedded software
嵌入式实时操作系统
embedded real-time operating system
确认测试　validation testing
软件　software
软件测试　software testing
软件设计　software design
生存周期　life-cycle model
时间邮票　time stamp
实时软件　real-time software
事务处理软件　transaction processing software
数据结构　data structure
第四代技术　fourth-generation technique
维护　maintenance
文档　document
系统软件　system software
演化　evolution
验收测试　acceptance testing
应用软件　application software
原型开发模型　prototype development model
指令　instruction
自补偿　self-compensating
自检　self-checking
自校准　self-calibration
自诊断　self diagnosis

1 Requirements, software designs and even test scripts have relatively short life expectancies compared to code.
需求、软件设计,甚至测试脚本与代码比较有相对短的预期寿命。

2 Modeling and analysis could become indispensable parts of process for software design and reengineering.
建模和分析可成为软件设计和重构过程的不可或缺的部分。

3 Since 1960s, a lot of software development methods have come out with the development of software engineering, such as waterfall model, spiral model, incremental model, etc.
自 20 世纪 60 年代以来,软件工程逐渐形成并发展,出现了很多软件开发模型与方法,例如瀑布模型、螺旋模型和增量模型等。

4 In the software design, with the top-down design ideas, hierarchical modular design is used in all the procedures to improve the portability of the software.
在软件设计中采用了自顶向下的设计思想,所有程序采用分层次模块化设计,提高了软件的可移植性。

第六节 可靠性与抗干扰技术
Section 6　Reliability and Anti-interference Technology

低温试验　low temperature test
电磁干扰　electromagnetic interference
对称电路　symmetrical circuit
浮置　floating
高温试验　high temperature test
隔离技术　isolation technique

第十九章 智能仪器设计

航空无线电公司分配法　method of aviation radio company distribution
机械试验　mechanical test
接收电路　receiving circuit
经济性　economy
均等分配法　method of equal distribution
看门狗　watch dog timer
抗干扰　anti-interference
抗干扰技术　anti-interference technology
可靠率　reliability rate
可靠性　reliability
可用性　availability
耦合通道　coupling channel
平均故障间隔时间　mean time between failures
平均修复时间　mean time to repair
屏蔽　shielding
冗余设计　redundancy design
软件可靠性设计　design of software reliability
软件陷阱　software trap
失效率　failure rate
时间模型　time model
数据模型　data model
消除干扰　interference elimination
抑制电磁干扰　restrain electromagnetic interference
硬件可靠性设计　design of hardware reliability
噪声源　noise source

1 During tests of the system, the anti-jam GPS overcame electronic jamming in various scenarios, including multiple simultaneous jamming.
在系统测试期间,抗干扰 GPS 克服了多种情况的电子干扰,包括多个同时发生的干扰。

2 It has accurate signal gathering, the design of specialized anti-electromagnetism interference, over electric flow over voltage protection, which can guarantee stability run of the system.
该技术有高精度信号采集,专门的抗电磁干扰设计,内部过流过压保护,确保系统的稳定可靠运行。

3 The proportional allocation factor and the allocation formula expressed in terms of failure rates and mean time between failures respectively are given.
给出了分别以失效率、平均故障间隔时间表达的分配比例因子及分配公式。

4 There were two ways to improve the availability, one of them is to decrease the mean time to repair (MTTR), the other is to increase the mean time between failures (MTBF).
主要从如何减少系统的平均修复时间(MTTR)和增加系统的平均无故障时间(MTBF)两方面来提高系统可用性。

第七节　可测试性设计
Section 7　Design for Testability

错误检测和校正码　error detection and correction code
故障隔离率　fault isolation rate
故障检测　fault detection
故障检测率　fault detection rate
机内测试　built-in test
机内测试设备　built-in test equipment
可测试性　testability
可测试性设计　design for testability
可观测性　observability
可控制性　controllability
可预见性　predictability
虚警率　false alarm rate
智能检测　intelligent detection
智能诊断　intelligent diagnosis

1 Design for testability is an important process in the chip design nowadays. Testability design of wireless chip needs a much higher requirement of test technology.
可测试性设计是现代芯片设计中的关键环节。无线接入芯片的可测试性设计对测试技术有更高的要求。

2 BIT is one important testability design of electronic systems. It's also one important way to test and insulate the failure in the electronic system or the internal equipment.
BIT(自检或机内测试)是电子系统可测试性设计的重要组成部分,是电子系统或设备内部检测和隔离故障的重要手段。

3 Design for software testability can increase the testability of software and, at the same time, reduce

the stress of software testing.
可测试性设计可以增强软件的可测试性,同时降低测试的强度。

4. The high false alarm rate is the main problem which puzzles the research and application of the BIT technology.
虚警率较高始终是困扰 BIT 技术研究和应用的主要问题之一。

5. For decreasing FAR, an improved model for BIT systems with embedded abnormal data detecting and renewing functions was proposed, and the effectiveness of the proposed model was proved.
为降低虚警率,提出嵌入数据异常检测和恢复功能的 BIT 系统改进模型,并证明了该模型的有效性。

第二十章 自动控制理论
Chapter 20 Automatic Control Theory

第一节 绪论
Section 1 Introduction

被控对象 controlled objective/plant	控制理论 control theory
被控制量 controlled variable	控制器 controller
比较机构 comparator	控制系统 control system
闭环控制系统 closed loop control system	控制信号 control signal
步进电动机 stepper motor	离散控制系统 discrete-time control system
参考输入 reference input	连续控制系统 continuous-time control system
测量元件 measure element	输出量 output variable
电流调节器 current regulator	速度调节器 speed regulator
定量的 quantitative	随机扰动 random disturbance
定性的 qualitative	位置随动系统 location-trace system
动态性能 dynamic property	稳定性 stability
反馈 feedback	稳定裕量 stability margin
反馈环节 feedback unit	稳态精度 steady-state accuracy
放大机构 amplifier	稳态误差 steady-state error
非线性控制系统 nonlinear control system	稳态值 steady-state value
干扰 disturbance	现代控制理论 modern control theory
给定量 input variable	线性控制系统 linear control system
环节 element/loop/unit	性能指标 performance index
经典控制理论 classic control theory	执行元件 actuator
开环控制系统 open loop control system	

In order to obtain more accurate control, the controlled signal must be fed back and compared with the reference input, and an actuating signal proportional to the difference of the output and the input must be sent through the system to correct the error. A system with one or more feedback paths like that just described is called a closed-loop system.

为了得到更加准确的控制,须将被控信号反馈到输入端与参考输入比较,得到一个与输出和输入的差值成比例的操作信号,送入系统从而校正误差。有一个或多个这样的反馈通道的系统称为闭环系统。

第二节 控制系统的数学模型
Section 2 Mathematical Model of Control Systems

比例环节 proportional element	惯性环节 inertial element
闭环传递函数 closed-loop transfer function	积分环节 integral element
并联 parallel connection	阶跃响应 step response
测速发电机 tacho-generator	开环传递函数 open-loop transfer function
传递函数 transfer function	零初始条件 zero initial condition
串联 series connection	脉冲响应 impulse response
等效变换 equivalent transform	梅逊公式 Mason formula
典型环节 nominal element	频率响应 frequency response
多回路系统 multiple-loop system	输入—输出描述 input-output description
根轨迹 root locus	数学模型 mathematical model

瞬态响应	transient response	信号流图	signal-flow graph
特征多项式	characteristic polynomial	引出点	leading-out point
微分方程	derivative equation	增益	gain
微分环节	derivative element	振荡环节	oscillatory element
系统框图	system block diagram	直流调速系统	DC speed control system
线性化	linearization	直流他励电动机	DC separately excited motor
相加点	summing junction	直流他励发电机	DC separately excited generator
斜坡响应	ramp response	滞后环节	delay element
谐波响应	harmonic wave response	状态变量	state variable

A summary of the properties of a transfer function is as follows:

(1) A transfer function is defined only for a linear system, and, strictly speaking, only for time-invariant systems.

(2) A transfer function between an input variable and an output variable of a system is defined as the ratio of the Laplace transform of the output to the Laplace transform of the input.

(3) All initial conditions of the system are assumed to be zero.

(4) A transfer function is independent of input excitation.

总结传递函数的性质如下：

(1) 传递函数只适用于线性系统，严格地讲，是线性时不变系统。

(2) 单输入单输出系统的传递函数定义为输出量的拉氏变换与输入量的拉氏变换之比。

(3) 系统的初始条件假设为零。

(4) 传递函数与输入激励无关。

第三节　控制系统的时域分析
Section 3　Time-domain Analysis of Control Systems

必要条件	necessary condition	临界稳定	marginally stable/critical stable
超调量	percentage overshoot	临界阻尼	critical damping
充分条件	sufficient condition	灵敏度	sensitivity
充要条件	sufficient and necessary condition	脉冲信号	impulse signal
单位阶跃响应	unit-step response	抛物线	parabola
低阶系统	lower-order system	前馈控制	feedforward control
调节时间	settling time	欠阻尼	underdamping
二阶系统	second-order system	上升时间	rise time
反变换	inverse transform	时域响应	time-domain response
峰值时间	peak time	实根	real root
负实部	negative real part	衰减	decay
复数根	complex root	稳态误差	steady-state error
高阶系统	high-order system	稳态响应	steady-state response
工程计算	engineering computation	无阻尼	undamp
共轭虚根	complex-conjugate root	无阻尼自然频率	undamp natural frequency
赫尔维兹判据	Hurwitz criterion	误差信号	error signal
加速度	acceleration	斜坡信号	ramp signal
渐近稳定	asymptotic stability	延迟时间	delay time
阶跃信号	step signal	一阶系统	first-order system
矩阵	matrix	右半平面	right-half plane
拉普拉斯变换	Laplace transform	正弦信号	sinusoidal signal
劳斯稳定判据	Routh's stability criterion	指令	command

终值定理　final-value theorem
重根　repeated roots
阻尼比　damping ratio

阻尼自然频率　damping natural frequency
最大超调量　maximum overshoot
左半平面　left-half plane

Routh-Hurwitz criterion: an algebraic method that provides information on the absolute stability of a linear time-invariant system. The criterion tests whether any roots of the characteristic equation lie in the right half of the s-plane. The number of roots that lies on the imaginary axis and in the right half of the s-plane is also indicated.

罗斯—赫尔维兹判据：判断线性时不变系统的绝对稳定性的一种代数方法。该判据判别特征方程式的根是否位于s平面右半平面，同时也可以判断位于s平面右半平面以及虚轴上的根的数目。

第四节　根轨迹
Section 4　Root Locus Method

参量　parameter
出射角　emergent angle
对称性　the symmetrical characteristic
非最小相位系统　nonminimum-phase system
分离点　breakaway point
复平面　complex plane
汇合点　break-in/convergence point
极点　pole
渐近线　asymptote
静态性能　static characteristic

开环极点　open-loop pole
开环增益　open-loop gain
零点　zero
起点　starting point
入射角　incident angle
实轴　real axis
虚轴　imaginary axis
终点　end point
主导极点　dominant pole

1 The root locus method was introduced by Evans in 1948 and has been developed and used extensively in control engineering practice. The root locus technique is a graphical method for sketching the locus of roots in the s-plane as a parameter is varied.

根轨迹方法是1948年由Evans首先引入的，经过逐渐发展被广泛地运用在控制工程实践中。根轨迹方法通常被认为是一种图解法，是闭环极点在s平面上随参数变化形成的轨迹图形。

2 The root locus will be drawn based on the following rules:
Rule 1. Draw the complex plane and mark the n open-loop poles and m zeros. The locus starts at a pole for K=0 and finishes at a zero or infinity when K=∞. The number of segments going to infinity is therefore n-m.
Rule 2. The loci are symmetrical about the real axis since complex roots are always in conjugate pairs. The angle between adjacent asymptotes is 360°/(n-m) and obeys the symmetry rule. The negative real axis is one asymptote when n-m is odd.

绘制根轨迹的基本原则：
规则1. 绘出复平面并标出n个开环极点和m个开环零点。根轨迹起始于K=0的极点，终止于K=∞的零点或无穷远。终点在无穷远的根轨迹为n-m条。
规则2. 因为复根均为共轭复数，所以根轨迹关于实轴对称。相邻渐近线的夹角为360°/(n-m)，同时遵守对称法则。当n-m为奇数时，负实轴为一条渐近线。

第五节　频率响应法
Section 5　Frequency Response Method

0型系统　type 0 system
1型系统　type 1 system
2型系统　type 2 system
闭合曲线　closed curve

伯德图　Bode diagram
穿越频率　crossover frequency
低频　low frequency
低通滤波器　low-pass filter

二阶因子	second-order element
幅频图	magnitude-frequency plot
复平面	complex plane
高频	high frequency
高通滤波器	high-pass filter
极坐标图	polar plot
剪切频率	sheared frequency
乃奎斯特图	Nyquist diagram
乃奎斯特稳定判据	Nyquist stability criterion
逆时针	counterclockwise
频带宽度	bandwidth
频域特性	frequency-domain specification
顺时针	clockwise
相对稳定性	relative stability
相频图	phase-frequency plot
相位	phase
相位交界频率	phase crossover frequency
相位裕量	phase margin
谐振峰值	resonance peak
谐振频率	resonance frequency
一阶因子	first-order element
有理数	rational
增益交界频率	gain crossover frequency
增益裕量	gain margin
振幅	magnitude
转折频率	break frequency

If the system is open-loop stable, then, for the closed-loop to be internally stable, it is necessary and sufficient that no unstable cancellations occur and that the Nyquist plot of $G_0(s)C(s)$ not encircle the point $(-1, 0)$.

If the system is open-loop unstable, with P poles in the open RHp, then, for the closed-loop to be internally stable, it is necessary and sufficient that no unstable cancellations occur and the Nyquist plot of $G_0(s)C(s)$ encircle the point $(-1, 0)$ P times counterclockwise.

If the Nyquist plot of $G_0(s)C(s)$ passes through the point $(-1, 0)$, there exists an $\omega_0 \in \mathbf{R}$ so that $F(j\omega_0)=0$, i.e. the closed-loop has poles located exactly on the imaginary axis. This situation is known as a critical stability condition.

如果系统是开环稳定的,则其闭环稳定的充分必要条件是没有不稳定抵消且 $G_0(s)C(s)$ 的乃奎斯特曲线不包括 $(-1,0)$ 点。

如果系统是开环不稳定的,有 P 个极点位于右半平面,则闭环系统稳定的充分必要条件是没有不稳定抵消且 $G_0(s)C(s)$ 乃奎斯特曲线按逆时针方向围绕 $(-1,0)$ 点旋转 P 周。

如果 $G_0(s)C(s)$ 乃奎斯特曲线通过 $(-1,0)$ 点,存在一个 $\omega_0 \in \mathbf{R}$ 使 $F(j\omega_0)=0$,即有闭环极点位于虚轴上,这种情况叫做临界稳定。

第六节 控制系统的校正
Section 6 Correction of Control Systems

补偿器	compensator
超前	lead
串联校正	cascade correction
反馈校正	feedback correction
幅值条件	amplitude condition
复合校正	compound correction
几何平均值	geometric mean
技术指标	technical index
经济性能	economic capability
静态误差系数	static error coefficient
可靠性	reliability
偶极子	doublet/dipole
前馈校正	feedforward correction
衰减特性	attenuation characteristic
图解法	diagram method
相角超前	phase lead
相角条件	phase condition
相角滞后	phase lag
校正	adjustment/correction
滞后	lag
最优控制	optimal control

So far, in this and the previous module, we have learnt two ways to compensate systems so as to achieve multiple performance requirements. In some cases such system just used as an example either method may be employed to achieve the objectives. If more performance measures are introduced, maybe only one of the techniques can be used. It is obvious that we can impose so many requirements that no compensation system will achieve the desired results, but the combined lead-lag compensation system method may allow more requirements to be met than by using either lead or lag compensation alone.

至此,在这个例子和前面的例子中,我们已经学习了两种校正系统使其全面满足性能要求的方法。在某些情况下,例如我们刚才用到的系统,两种校正法都可用。但是如果更多的性能要求引入,可能就会只有一种方法可用。显然如果要求的性能指标太多,那么两种方法则都无法满足要求,但是组合的超前-滞后校正则比单独的超前校正或滞后校正能够更好地满足系统多方面的要求。

第七节　PID 控制与鲁棒控制
Section 7　PID Control and Robust Control

PD 控制　proportional-plus-derivative control
PID 控制　proportional-plus-integral-plus-derivative control
PI 控制　proportional-plus-integral control
标称传递函数　nominal transfer function
乘法摄动　multiplicative perturbation
传感器　sensor
跟踪误差　tracking error
加法摄动　addition perturbation
解析法　analysis method
临界增益　critical gain
鲁棒性　robustness
模型误差　modeling error
齐格勒—尼可尔斯法则　Ziegler-Nichols method
摄动　perturbation
双重零点　dual zeroes
线性时不变　linear time-invariant
噪声信号　noise signal
自由度　freedom

Ziegler-Nichols (Z-N) Oscillation method:
This procedure is valid only for open-loop stable plant, and it is carried out by means of the following steps.
(1) Set the true plant under proportional control with a very small gain.
(2) Increase the gain until the loop starts oscillating.
(3) Record the controller critical gain $K_p = K_c$ and the oscillation period of the controller output, T_c.
(4) Adjust the controller parameters.

齐格勒—尼可尔斯(Z-N)振荡法则:
该法则只适用于开环稳定系统,按如下步骤操作:
(1)使真实对象只有比例控制,并有较小的增益。
(2)逐渐增加增益值使系统开始振荡。
(3)记录此时控制器的临界增益 $K_p = K_c$ 以及控制器输出的振荡周期 T_c。
(4)确定控制器参数。

第八节　离散控制系统
Section 8　Discrete-time Control Systems

部分分式　partial fraction
采样控制系统　sampled-data control system
差分方程　differential equation
次要带　secondary domain
调制器　modulator
迭代法　iterative method
分辨率　resolution ratio
分时处理　time-sharing processing
复现　reconstruction
复相　complex phase
级数求和　summation of series
卷积定理　convolution theorem
留数　residue
脉冲传递函数　pulse-transfer function
脉冲序列　pulse sequence
模拟　analog
数码序列　digital sequence
数字控制器　digital controller
微商　derivative
位移定理　displacement theorem
香农定理　Shannon theorem
载波信号　carrier signal
主要带　domain

A very valuable and frequently used tool for plant modeling is a state-variable description. State variables come from a set of inner variables which is a complete set, in the sense that, if these variables are known at some time, then any plant output, y(t), can be computed, at all future times, as a function of the state variables and the present and future values of the inputs.

状态变量描述是系统建模的一种非常有用而且很常用的方法。状态变量由全体内变量组成,从某种意义上讲,这些变量是一种集合,如果这些变量在某一时刻是已知的,那么在任何未来时刻的对象输出 y(t)就都可以作为状态变量以及输入的现在与未来值的函数而计算出来。

第九节 状态空间分析法
Section 9 State Space Analysis

初始状态　initial state
存在性定理　existence theorem
动力学系统　dynamic system
对角标准形　diagonal canonical form
对角化　diagonalization
对偶　antithesis
多项式　polynomial
非负定　nonnegative definite
负定　negative definite
极点配置　pole placement
可观测的　observable
可控的　controllable
控制矩阵　control matrix
李亚普诺夫方程　Lyapunov function
李亚普诺夫稳定性判据　Lyapunov stability criterion
列向量　column vector
能观性　observability
能观性矩阵　observability matrix
能控标准形　controllable canonical form
能控性　controllability
能控性矩阵　controllability matrix
偶数　even
耦合　coupling
齐次方程　homogeneous equation
奇数　odd

奇异矩阵　singular matrix
塞尔维斯特　Sylvester
输出方程　output equation
特征方程　characteristic equation
特征向量　eigenvector/proper vector
特征值　eigenvalue/characteristic value
维数　dimension
线性时不变系统　linear time-invariant system
行列式　determinant
行向量　row vector
雅可比矩阵　Jacobian matrix
约当标准形　Jordan canonical form
正定　positive definite
指数函数　exponential function
秩　rank
主对角线　leading diagonal
状态　state
状态变量　state variable
状态方程　state equation
状态估计　state estimate
状态观测器　state observer
状态空间　state space
状态矢量　state vector
状态转移矩阵　state transition matrix

It is also possible to determine whether a system is controllable and observable directly from the original system equations

$$\dot{x} = Ax + Bu$$
$$y = Cx + Du$$

To determine controllability, we form the controllability matrix S_c from

$$S_c = [B \quad AB \quad A^2B \quad \cdots \quad A^{n-1}B]$$

The system is controllable if the rank of S_c is n, the order of the system. Similarly, from original system equations we may form the observability matrix S_0 from

$$S_0 = \begin{bmatrix} C \\ CG \\ CG^2 \\ \vdots \\ CG^{n-1} \end{bmatrix}$$

The test for observability is that the rank of S_0 must also be n, the system order.

也可以由系统的原始状态方程直接判断系统的能控性和能观性。

$$\dot{x} = Ax + Bu$$
$$y = Cx + Du$$

判断能控性，列写出能控性矩阵 S_c 为

$$S_c = [A \quad AB \quad A^2B \quad \cdots \quad A^{n-1}B]$$

当 S_c 的秩为 n，即系统阶数时，系统的状态是完全能控的。同样，根据原始状态方程列写出能观性矩阵 S_0 为：

$$S_0 = \begin{bmatrix} C \\ CG \\ CG^2 \\ \vdots \\ CG^{n-1} \end{bmatrix}$$

当 S_0 的秩为 n，即系统阶数时，系统的状态是完全能观的。

第十节 非线性控制系统
Section 10 Nonlinear Control Systems

鞍点　saddle point
饱和　saturation
参数优化　parametric optimization
发散　divergence
非线性滤波器　nonlinear filter
畸变　distortion/skewness
极限环　limit cycle
继电器特性　relay characteristic
焦点　focus/focal point
螺旋线　spiral/helical
描述函数　describing function
描述函数法　describing function method

收敛　convergence
死区　dead zone
泰勒级数　Taylor series
图解法　diagram method
线性化　linearization
相平面法　the phase plane method
谐波分量　harmonic wave component
因变量　dependent variable
中心点　center point
自变量　independent variable
自持振荡　self-maintained oscillation

Lyapunov stability criteria:

If a system possesses a Lyapunov function so that $\dot{V}(x)$ is negative definite or negative semidefinite along the trajectories of the system (trajectory approaches origin when $V(x) < 0$), then the origin is said to be asymptotically stable. For this case, equilibrium is obtained at the origin.

If a system possesses a Lyapunov function so that $V(x) > 0$ is positive definite or positive semidefinite along the trajectories of the system, then the origin is said to be unstable. For this case, the trajectories move away from the origin.

If a system possesses a Lyapunov function so that $V(x)$ remains zero, the trajectory remains on a path of constant $V(x)$. Such a trajectory is called a limit cycle.

If $V(x)$ is indefinite, the test fails, as neither stability nor instability has been proved. Thus, one must try a different V(x) function. A system may have many Lyapunov functions. However, if one has been found to prove that the origin is unstable, then it is unstable. The major obstacle in applying Lyapunov's method is that of obtaining a satisfactory Lyapunov function to prove stability or instability, as the case may be.

李亚普诺夫稳定性判定：

如果系统具有李亚普诺夫函数 V(x) 是负数或准负数并沿着系统的轨迹（当 V(x) 小于 0 的时候轨迹逼近原点）运行,那么原系统是趋于稳定的,对于这种情况,系统在原点获得平衡。

如果系统具有李亚普诺夫函数 V(x) 大于 0,且为正数或准正数沿着系统的轨迹运行,则原系统被认为是不稳定的,这种情况下,轨迹远离原点。

如果系统具有李亚普诺夫函数 V(x) 等于 0,那么轨线保持在常量 V(x) 的轨道上,这样的轨迹被称为极限循环。

如果 V(x) 是不定的,如果系统被证实既非稳定也非不稳定,则测试失败。因而,必须尝试采用不同的 V(x) 函数。一个系统可以拥有多个李亚普诺夫函数。如果其中一种函数已经被发现证实系统是不稳定的,那么这个系统就是不稳定的。在应用李亚普诺夫的方法的主要障碍就是如何获得一个令人满意的李亚普诺夫函数来证明系统稳定或不稳定,这需要视具体情况而定。

第二十一章　数字电子技术
Chapter 21　Digital Electronics Technology

第一节　绪论
Section 1　Introduction

比特率　bit rate
调幅　amplitude modulation
调频　frequency modulation
二进制编码的十进制数　binary-coded-decimals(BCD)
二值数字逻辑　binary digital logic
反码　one's complement
符号数　signed number
负号　minus sign
格雷码　gray code
互补金属氧化物半导体门电路　complementary metal-oxide-semiconductor (CMOS)
基数　radix
晶体管-晶体管逻辑　transistor-transistor logic (TTL)
离散信号　discrete signal

连续信号　continuous signal
量化　quantification
逻辑电平　logic level
脉冲重复频率　pulse repetition rate (PRR)
模拟信号　analog signal
权　weight
上升时间　rise time
时序图　timing diagram
无符号数　unsigned number
下降时间　fall time
溢出　overflow
余3码　excess three code
正号　plus sign
周期性的　seasonal
最低位　least significant digit (LSD)
最高位　most significant digit (MSD)

1 Electronic circuits can be divided into two broad categories: digital and analog. Digital electronics involves quantities with discrete values, and analog electronics involves quantities with continuous values. Although you will be studying digital fundamentals in this book, you should also know about analog because many applications require both.

电路可分为两大类——数字电路和模拟电路。数字电路包括以离散值表示的量,模拟电路则包括连续值表示的量。尽管在这里我们将学习数字量的基本原理,你也需要知道模拟量的知识,因为许多电路两者都会用得上。

2 Decimal to binary conversion: Divide the decimal number to be converted successively by 2, ignoring the remainders, until you have a quotient of 0. The remainders will be used later to determine the answer.

十进制转换成二进制的方法:将要转换的十进制数连续地被2除,每一步的余数先不考虑,直到商为0,然后可以根据每一步得到的余数求出二进制数。

第二节　逻辑代数基础
Section 2　Logic Algebra Foundation

布尔代数　Boolean algebra
乘积项　product term
对偶函数　dual function
反函数　complement function
反演规则　complementary operation theorem
非　no
分配律　distributive theorem
恒等式　identity

化简　simplification
或　or
或非　nor
交换律　commutative theorem
结合律　associative theorem
竞争冒险　race and hazard
卡诺图　Karnaugh map
逻辑变量　logic variable

逻辑表达式　logic expression
逻辑常量　logic constant
逻辑符号　logic symbol
逻辑函数　logic function
摩根定理　De Morgan's theorem
偶校验　even check
奇偶发生/检验器　parity generation /checker
奇校验　odd check
特性方程　characteristic equation
同或　exclusive-nor/nonexclusive or

无关项　don't care terms
相邻项　adjacency
异或　exclusive-or
与　and
与非　nand
与或非　and-or-invert
真值表　truth table
组合逻辑　combinational logic
最大项　maxterm
最小项　miniterm

1 Boolean algebra which was formulated by George Boole, an English mathematician (1815-1864), described propositions whose outcome would be either true or false. In computer work it is used in addition to describe circuits whose state can be either 1 (true) or 0 (false).

布尔代数由英国数学家乔治·布尔(1815—1864)创立,用来描述结果只有真或假两种结果的命题。在计算机中,它还用来描述电路的两种状态1(真)或0(假)。

2 A product term in which all the variables exactly appear once, either complemented or uncomplemented, is called a miniterm. Its characteristic property is that it represents exactly one combination of the binary variables in a truth table. It has the value 1 for that combination and 0 for all others.

在乘积项中,所有的变量都以原变量或反变量的形式出现一次,且只出现一次,则该乘积项称作最小项。最小项的特点是它只与变量的一组取值组合相对应,只有该组取值才使其为1,其他的组合取值都为0。

第三节　逻辑门电路
Section 3　Logic Gate Circuit

饱和电流　saturation current
传输门　transmission gate(TG)
传输特性　transfer characteristic
传输延迟时间　propagation delay time
动态特性　dynamic characteristic
多发射极三极管　multiemitter transistor
多级与非门电路　multilevel nand circuit
二极管　diode
二输入与门　two-input and gate
反相器　inverter
负逻辑　negative logic
干扰信号　disturbance signal
高阻态　high impedance state
工作点　operating point
功耗　power dissipation
集电极开路　open collector
开关时间　switching time
开关特性　switching characteristic
开启电压　threshold voltage
拉电流　draw-off current

临界　critical
门阵列　gate array
模拟开关　analog switch
片选　chip select
三极管　bipolar junction transistor(BJT)
三极管—三极管电路　transistor-transistor logic(TTL)
三态门　three-state gate
扇出　fan out
扇入　fan in
上拉电阻　pull-up resister
射极耦合　emitter-coupled
双列直插式封装　dual in-line package (DIP)
图腾柱　totem pole
线与　wire-and
肖特基二极管　Schottky diode
延时—功耗积　time delay-power dissipation product
有源下拉电路　active pull-down circuit
噪声容限　noise margin
正逻辑　positive logic

1 The three-state, tri-state, or 3-state logic allows output ports to assume a high impedance state in addition to the fundamental 0- and 1-levels, effectively "removing" the output from the circuit. This allows multiple circuits to share the same output line or lines (such as a bus).

三态或三态逻辑，允许输出端除了输出 0 状态和 1 状态之外，还可输出一个高阻态。这个高阻态可以有效地将输出从电路中"移除"。这就允许多个电路共享一条或多条输出线（类似总线）。

2 Positive and negative logic：
Except during transition, the binary signals at the inputs and outputs of any gate have one of two values：H or L. One value represents logic 1 and the other logic 0. Choosing the high level H to represent logic 1 defines a positive-logic-system, and choosing the low level L to represent logic 1 defines a negative-logic-system.

正逻辑和负逻辑：
除了在传输的过程中，一个二进制信号在门电路的输入和输出端只有两种取值中的一个：高电平或低电平，一个代表逻辑 1，另一个代表逻辑 0。如果用高电平代表逻辑 1，则是正逻辑系统，如果用低电平代表逻辑 1，则是负逻辑系统。

第四节　基础组合逻辑电路
Section 4　Combinational Logic Circuit

半加器	half adder	借位	borrow
被加数	augend	进位	carry
被减数	minuend	扩展	expansion
编码	encoding	来自低位进位	carry-in
编码器	encoder	码制转换器	code converter
并行加法器	parallel adder	七段显示器	seven-segment display
差	difference	全加器	full adder
超大规模集成电路	very large-scale integrated(VLSI)	使能端	enable point
超前进位产生器	look-ahead carry generator	数码比较器	digital comparator
超前进位加发器	look-ahead carry adder	数码显示器	digital display
串行进位加法器	ripple carry adder	算术逻辑单元	arithmetic logic unit(ALU)
大规模集成电路	large-scale integrated(LSI)	向高位进位	carry-out
多路数据分配器	multiplexer	小规模集成电路	small-scale integrated (SSI)
多路数据选择器	demultiplexer	液晶显示器	liquid crystal display(LCD)
二—十进制译码器	BCD decoder	译码器	decoder
功能表	function table	荧光数码管	fluorescent nixie tube
管脚	pin	优先编码器	priority encoder
函数产生器	function generator	约束	inhibit
辉光数码管	glow discharge nixie tube	中规模集成电路	medium-scale integrated (MSI)
加数	addend	专用集成电路	application specific integrated circuit
减法器	subtractor		
减数	subtrahend		

1 A decoder is a combinational circuit that covers binary information from the n coded inputs to a maximum of 2^n unique outputs. If the n-bit coded information has used bit combinations, the decoder may have fewer than 2^n output.

译码器是能将 n 位二进制数表示的特定的信息"翻译"出来的一种组合电路。全译码的译码器有 2^n 个输出端，一个输出端对应一个最小项。如果 n 位编码的信息用到了少量组合，则输出端的个数会小于 2^n。

2 Encoder: An encoder is a combinational logic circuit that essentially performs a "reverse" decoder function. An encoder accepts an active level on one of its inputs representing a digit, such as a decimal or octal digit, and converts it to a coded output, such as BCD or binary.

编码器：编码器基本上是一个与译码器函数相反的组合电路。当编码器的一个输入端输入有效电平时，它代表一个数字，可以是十进制数也可以是八进制数，然后编码器把它转换成二-十进制或二进制代码输出。

第五节　锁存器和触发器
Section 5　Latch and Flip-Flop

D 锁存器	D latch	上升沿	rise edge
保持时间	hold time	时序电路	sequential circuit
初始状态	initial state	时钟触发器	clocked flip-flop
电平触发	level trigger	时钟信号	clock signal
抖动	dithering	特性表	characteristic table
翻转	overturn	同步触发器	synchronous flip-flop
负边沿	negative edge	图形符号	graphic symbol
复位	reset	未定义	undefined
机械开关	mechanical on-off	下降沿	fall edge
基本 RS 触发器	basic RS flip-flop	延迟	postpone
可能状态	potential state	正边沿	positive edge
脉冲	pulse	置位	set
清零	clear	主从触发器	master-slave flip-flop
驱动方程	driving equation		

1 D latch is an electronic device that can be used to store one bit of information. The D latch is used to capture, or "latch" the logic level which is presented on the data line when the clock input is high. If the data on the D line changes state while the clock pulse is high, then the output changes follows the input.

D 锁存器是一种能用来存储一位信息的电子设备。当时钟输入时高电平时，D 锁存器用来捕捉或"锁存"数据线上的逻辑电平值。在时钟脉冲高电平期间，如果数据线上的数据发生了变化，那么输出端也会随着输入的变化而变化。

2 The Master-Slave Flip-Flop consists of two grade Flip-Flops. One receive the input signals and its state is determined from input, is called the Master; the other, whose input is linked up with Master's output and whose state is determined from Master's state, is called the Slave.

主从触发器由两级触发器构成，其中一个接收输入信号，它的状态直接受输入的控制，称为主触发器。另一个的输入端与主触发器的输出端相连，它的状态受主触发器的控制，称为从触发器。

第六节　时序逻辑电路
Section 6　Sequential Logic Circuit

保持、有效	hold	二进制计数器	binary counter
闭合回路	closed circuit	翻转触发器	toggle flip-flop
并入串出	parallel-in serial-out	反馈	feedback
串行输入	serial input	分频	frequency division
次态	next state	环形计数器	orbicular counter
存储电路	memory circuit	激励表	excitation table
等价状态	equivalence state	计数模式	counter mode
递增-递减计数器	up-down counter	可逆计数器	reversible counter

模　module
频率计　frequency counter
启动　startup
手册、指南　manual
双向移位寄存器　bidirectional shift register
同步时序电路　synchronous sequential circuit
无效状态　invalid state
现态　present state
移位寄存器　shift register
异步时序电路　asynchronous sequential circuit
状态表　state table
状态化简　state simplification
状态图　state diagram
状态转换　state conversion
总线　bus

1 The behavior of a sequential circuit is determined from the input, output, and present state of the circuit. The output and the next state are a function of the inputs and the present state.

时序逻辑电路的行为不仅仅受输入信号的影响，还受电路的输出信号和现态的控制。其输出信号和次态是输入信号和现态的函数。

2 A register include a set of flip-flops. Since each flip-flop is capable of storing one bit of information, an n-bit register, including n flip-flops, is capable of storing n-bit of binary information. By the broadest definition, a register consists of a set of flip-flop, together with gates that implement their state.

寄存器包括一组触发器。因为一个触发器能存储一位二进制信息，所以 n 位寄存器包含 n 个触发器，能存储 n 位二进制信息。广义上讲，寄存器由触发器和门电路组成，它们一起决定寄存器的状态。

第七节　脉冲波形的产生与整形
Section 7　Pulse Waveform Generation and Transformation

555 定时器　555 timer
不可重复触发　nonretriggerable
充电　charge
单稳态触发　monostable multivibrator
电压比较器　voltage comparator
多谐振荡器　astable multivibrator
放电　discharge
回差电压　backlash voltage
可重复触发　retriggerable
脉冲宽度　pulse stretcher
上限阈值电压　upper threshold
施密特触发器　Schmitt trigger
石英晶体振荡器　crystal oscillator
双稳态　bistable
微分型　differential
下限阈值电压　lower threshold
阈值电平　threshold level
振荡周期　astable period
滞后　hysteresis

1 The output of a monostable multivibrator presents two states, one of which is stable, while the other is unstable. A trigger causes the circuit to enter the unstable state. After entering the unstable state, the circuit will return to the stable state after a set time. Such a circuit is useful for creating a timing period of fixed duration in response to some external event.

单稳态触发器的两个输出状态中一个状态是稳定的，而另一个状态是不稳定的。一个触发可引起电路进入不稳定状态。进入不稳定状态后，在设定的时间内，电路又会返回稳定状态。这种电路通常用来设定一段固定的时间用来响应外部事件。

2 Schmitt trigger is a special trigger that has backlash voltage. As the input voltage rises, the output stays at a low or 0 value until the input voltage reaches upper threshold. At this upper threshold, the output snaps to a logic 1 value. When the input voltage drops, the output does not return to logic 0 until the input voltage drops below the lower threshold. The difference in the upper threshold and lower threshold is called backlash voltage.

施密特触发器是一种具有回差电压的特殊触发器。输入电压在上升的过程中，触发器输出一直保持低电平不变，直到上升到高电平阈值电压，这时，输出从低电平跳到高电平。输入电压在下降的过程中,下降到低电平阈值电压时,触发器的输出由高电平下降到低电平。高电平阈值电压和低电平阈值电压之间的差值,称为回差电压。

第八节 半导体存储器
Section 8　Semiconductor Memory

存储单元　memory cell
存储矩阵　memory array
存储时间　storage time
电可擦除的 PROM　electricity EPROM (E^2 PROM)
动态存储器　dynamic memory
读/写存储器　read-write memory (RWM)
读周期　read period
静态存储器　static memory
可编程 ROM　programmable ROM (PROM)

可编程阵列逻辑　programmable array logic(PAL)
可擦可编 ROM　erasable programmable ROM (EPROM)
快闪存储器　flash memory
列选择线　column-select line
随机存储器　random access memory (RAM)
行选择线　row-select line
再生(刷新)　regenerate (refresh)

1 Read-only memory (ROM) is a class of storage medium used in computers and other electronic devices. Data stored in ROM cannot be modified, or can be modified only slowly or with difficulty, so it is mainly used to distribute firmware (software that is very closely tied to specific hardware, and unlikely to need frequent updates).

只读存储器(ROM)是一类用在计算机或电子设备中的存储媒介。存储在 ROM 中的数据不能被修改,或者修改起来非常慢,或者非常困难,所以它主要用来放置固定程序(与硬件电路联系很紧密的一种程序软件,不需要频繁更新)。

2 Random access memory (RAM) is one of typical memories that are used in various parts of a computer. RAM accepts new information for storage to be available later for use. The process of storing new information in memory is referred to as a memory write operation. The process of transferring the stored information out of memory is referred to as a memory read operation.

随机存储器(RAM)是组成计算机的基本存储单元之一。它接收新的信息并存储起来以备后用。存储器接收新信息的过程称为"写"操作。将存储的信息输出的过程称为"读"操作。

第九节 可编程逻辑器件
Section 9　Programmable Logic Device(PLD)

布线　wiring
擦除　erase
复杂可编程逻辑器件　complex programmable logic device(CPLD)
宏单元　macro cell
集总布线区　global routing pool(GRP)
可配置逻辑块　configurable logic block (CLB)
逻辑单元阵列　logic cell array (LCA)
内部连线　interconnect
输出逻辑宏单元　output logic macro cell (OLMC)

输出输入块　input output block (IOB)
通用逻辑块　generic logic block(GLB)
通用阵列逻辑器件　generic array logic (GAL)
现场可编程逻辑阵列　field programmable logic array (FPLA)
现场可编程门阵列　field programmable gate array (FPGA)
硬链接　hard link
在系统可编程　in-system programmability(ISP)
知识产权　intellectual property(IP)

1 A programmable logic device or PLD is an electronic component used to build reconfigurable digital circuits. Unlike a logic gate, which has a fixed function, a PLD has an undefined function at the time of manufacture. Before the PLD can be used in a circuit it must be programmed, that is, reconfigured.

第二十一章 数字电子技术

可编程逻辑器件是用来重构数字电路的电器组件。与逻辑门电路不一样,每个逻辑门电路都有它固定的逻辑功能,而 PLD 在生产的时候并没有固定的功能。想要在电路中使用 PLD,之前必须要对它进行编程,这就是所谓的"重构"。

2 A field-programmable gate array (FPGA) is an integrated circuit designed to be configured by a customer or a designer after manufacturing——hence "field-programmable". The FPGA configuration is generally specified using a hardware description language (HDL), similar to that used for an application-specific integrated circuit (ASIC).

现场可编程逻辑阵列(FPGA)是一种集成电路,在生产出来之后可由用户或设计师自行配置使用,因此称为现场可编程。FPGA 的配置一般使用硬件描述语言(HDL),这种语言类似于一种为专用集成电路(ASIC)而设计的语言。

第十节 数模与模数转转器
Section 10 Digital-to-Analog and Analog-to-Digital Converter

采样率　sampling rate　　　　　　取样—保持电路　sample-hold circuit
单调性　monotonicity　　　　　　　权电流　weighted current
倒置梯形　inverted ladder　　　　　权电阻　weighted resistor
分辨率　resolution　　　　　　　　双积分　dual integration
精度　precision　　　　　　　　　　逐次逼近　successive approximation
量化误差　quantization error

1 In electronics, a digital-to-analog converter (DAC or D-to-A) is a device that converts a digital (usually binary) code to an analog signal (current, voltage, or electric charge). An analog-to-digital converter (ADC) performs the reverse operation. Signals are easily stored and transmitted in digital form, but a DAC is needed for the signal to be recognized by human senses or other non-digital systems.

在电子学中,数模转换器(DAC)是一种可以将数字量(通常是二进制)转换成模拟量(电流、电压或电荷)的设备。模数转换器(ADC)执行的是相反的过程。信号以数字量的形式易于存储和传输,但是 DAC 对于那些人类感官和非数字系统要处理的信号来说就是必要的了。

2 Precision is the maximal windage of convertor's actual conversion characteristic curve and perfect conversion characteristic curve. Convertor's precision is determined from various errors. Three basic error sources are: maladjustment error, plus error and nonlinear error.

精度是转换器实际转换特性曲线与理想转换特性曲线之间的最大偏差。转换器的转换精度由各项误差综合决定。转换器有三种基本误差源:失调误差、增益误差、和非线性误差。

第二十二章 模拟电子技术
Chapter 22　Analog Electronics Technology

第一节　绪论
Section 1　Introduction

波动　fluctuation
波形整形　wave-shaping
采样　sample
参数　parameter
操纵　manipulate
初始电平　the original level
导航　navigation
电流放大模型　the current-amplifier model
电流增益　current gain
电压放大模型　the voltage-amplifier model
电源　power supply
电源效率　power efficiency
分贝表示法　decibel notation
分流电路　discrete circuit
峰值检波器　the peak detector
负载效应　loading effect
附加噪声　the addition of noise
复数增益　complex gain
干扰　disturbance

功率增益　power gain
功能块　functional block
互阻放大模型　the transresistance-amplifier model
激励　stimulus
级联放大器　cascaded amplifier
集成电路　integrated circuit
量化误差　quantization error
临界　extreme
流程图　flowchart
输入阻抗　input impedance
提取　extract
信号源　signal source
振幅　amplitude
整流器　rectifier
整体设计过程　the overall process
子系统　subsystem
组件　component
最大限度提高性能　maximize performance

Amplifiers can be modeled as voltage amplifiers, current amplifiers, transresistance amplifiers and transconductance amplifiers.
放大器可以分成电压放大器、电流放大器、互阻放大器和跨导放大器。

第二节　运算放大器
Section 2　Operating Amplifier

T型反馈网络　T type feedback network
巴特沃思滤波器　Butterworth filter
比例运算电路　scaling circuit
差分输入　differential input
乘法运算电路　multiplication circuit
除法运算电路　division circuit
电压跟随器　voltage follower
对数运算电路　logarithmic circuit
积分运算电路　integration circuit
求和电路　summing circuit

同相输入　non-inverting input
外特性　external characteristic
微分电路　differential circuit
虚地　virtual ground
虚短电路　virtual short circuit
运算电路　operational circuit
指数电路　exponential circuit
中心频率　center frequency
状态变量　state variable

　　If a Gaussian process is stationary, then the process is also strictly stationary. If a Gaussian process X(t) is applied to a stable linear filter, then the random process Y(t) developed at the output of the filter is also Gaussian. Gaussian processes that are stationary in the narrow sense may be realized by way of certain dynamical systems.

第二十二章 模拟电子技术

如果高斯过程是平稳的,那么整个过程也一定是严格平稳的。如果高斯过程 X(t) 通过一个稳定线性滤波器,那么,滤波器的输出 Y(t) 也是高斯型的。平稳高斯过程从狭义上来说可以通过动态系统来实现。

第三节 二极管及其基本电路
Section 3 Diodes and Diode Circuits

N 型　N type
P 型　P type
半导体　semiconductor
半导体二极管　semiconductors diode
半导体基本概念　basic semiconductor concept
本征　intrinsic
参数　parameter
掺杂　doping
等效电路　equivalent circuit
点接触型　point contact type
电流方程　current equation
二极管　diode
发光　light-emitting
放大区　active region
伏安特性　volt ampere characteristic
符号　symbol
共价键　covalent bond
光电二极管　photodiode
耗尽层　depletion region/depletion zone
恒流区　constant current region
击穿电压　breakdown voltage
极间电容　interelectrode capacitance
结构　construction

结型　junction type
截止区　cutoff region
开关高频二极管的性能　switching and high-frequency behavior
开启电压　threshold voltage
空穴　hole
扩散电容　diffusion capacitance
理想二极管模型　the ideal diode model
理想模型　ideal model
面结型二极管的物理机理　physics of the junction diode
势垒电容　barrier capacitance
输出特性　output characteristic
稳压电路　voltage-regulator circuit
稳压二极管　zener diode
线性小信号等效电路　linear small signal equivalent circuit
小信号模型　small signal model
信噪比　signal-to-noise ratio
载流子　carrier
折线化模型　piecewise model
整流器　rectifier
自由电子　free electron

1 U_D and i_D represent the total instantaneous diode voltage and current. At times, we may wish to emphasize the time-varying nature of these quantities and then we use v-p(t) and i-p(t).
u_D 和 i_D 表示二极管的电压和电流总瞬时二值。有时,为了强调这些量随时间变化的性质,我们用 v-p(t) 和 i-p(t) 表示瞬时值。

2 V_Q and I_Q represent the DC diode current and voltage at the quiescent point.
V_Q 和 I_Q 表示二极管在静态工作点处的直流电压和电流。

3 Diodes that are intended to operate in the breakdown region are called zener diodes. Zener diodes are often used in applications for which a constant voltage in breakdown is desirable.
工作在击穿区域的二极管被称为齐纳二极管。齐纳二极管最可取的是它在反向击穿状态下的恒压特性。

4 The ideal diode model is a short circuit for forward currents and an open circuit for reverse voltages.
理想二极管模型是加正向电压短路有正向电流,加反向电压开路。

5 In the forward direction, the ideal diode conducts any current forced by the external circuit while displaying a zero voltage drop. The ideal diode does not conduct in the reverse direction; any applied voltage appears as reverse bias across the diode.
理想二极管在正向偏置时导通,二极管两端的压降为零;反向偏置时截止,二极管反向偏置电压为外加电压。

第四节 双极结型三极管及放大电路基础
Section 4 Bipolar Junction Transistors and Amplifier Base

h参数模型　h-parameter model
h参数小信号模型　hybrid small signal model
U_{BE}倍增　U_{BE} multiplier
饱和区　saturation region
变跨导型乘法器　variable transconductance multiplier
变压器耦合　transformer coupled
变压器耦合放大电路　transformer coupled amplifying circuit
波特图　Bode diagram / Bode plot
超前校正　phase lead compensation
电流放大系数　current amplification factor
多发射结晶体管　multiple emitter transistor
多级放大电路　multistage amplifying circuit
发射结电容　emitter junction capacitance
反向饱和电流　reverse saturation current
反向偏置　reverse bias
放大电路　amplifying circuit
放大概念　amplification concept
放大区　amplification region
放大作用　amplification effect
非线性失真　non-linear distortion
幅值裕度　gain margin
负载线　load line
负载线分析　load line analysis
工作点稳定　operating point stabilization
共基组态　common-base configuration
共集组态　common-collector configuration
共射截止频率　common-emitter cutoff frequency
共射组态　common-emitter configuration
光电耦合器　photocoupler
集电结电容　collector junction capacitance
交流负载线　alternating current load line
交流通路　alternating current path
截止区　cutoff region
静态工作点　quiescent point
连接方式　configuration
零点漂移　zero drift
耦合电容　coupling capacitance
旁路电容　by-pass capacitance
偏置电流　bias current
频率特性　frequency characteristic
频率响应　frequency response
上限截止频率　upper cut-off frequency / high 3-dB frequency
射级偏置电路　emitter biasing circuit
射极跟随器　emitter follower
输出电阻　output resistance
输入电阻　input resistance
特性曲线　characteristic curve
特征频率　characteristic frequency
通频带　pass band
图解分析法　graphical analysis
稳定判据　stability criterion
稳定裕度　stability margin
下限截止频率　lower cut-off frequency / low 3-dB frequency
相频特性　phase frequency characteristic
相位裕度　phase margin
校正措施　compensation method
有源元件　active component
正向偏置　forward bias
直接耦合放大电路　direct coupled amplifying circuit
直流模型　DC model
滞后校正　phase lag compensation
中频　intermediate frequency
转移特性　transfer characteristic
假互补电路　quasi complementary circuit
阻容耦合放大电路　RC coupling amplifier
最大输出幅值　maximum output amplitude

1 Depending on the bias conditions on its two junctions, the BJT can operate in one of three possible modes: cutoff (both junctions reverse-biased), active (the EBJ forward-biased and the CBJ reverse-biased), and saturation (both junctions forward-biased). For amplifier applications, the BJT is operated in the active mode. Switching applications make use of the cutoff and saturation modes.

根据两结的不同偏止状态,BJT 在三种状态之一下工作:两结反偏—截止;发射结正偏,集电结反偏—放大;两结正偏—饱和。在放大电路中,BJT 工作在放大状态,在开关电路中,BJT 工作在截止和饱和状态。

2 Two amplifiers are cascaded by connecting the input of the second to the output of the first. The overall voltage, current, or power gains are the product of the respective individual gains, taking loading of the first stage by the second into account.
两个放大器级联就是第二级的输入连接第一的输出。总的电压、电流或功率增益是各级增益的乘积，前级的负载取决于第二级的输入。

第五节 场效应管放大电路
Section 5　Field-effect Amplifier

CMOS 电路　CMOS circuit
衬底　substrate
单极型晶体管　unipolar transistor
多晶硅　polysilicon
恒流区　constant current region
夹断电压　pinch off voltage
结型　junction type
金属氧化物半导体场效应管　metal oxide semiconductor FET
绝缘栅型　isolated gate type
开启电压　threshold voltage
漏极　drain
源极　source
增强型　enhancement
栅极　gate
栅极偏置电路　gate bias circuit

1 Distortion occurs in FET amplifiers because of the nonuniform spacing of the drain characteristics. Distortion is less pronounced for small signal amplitudes.
FET 放大器发生失真的原因是漏极特性间距的不均匀。信号幅度小时失真不明显。

2 For use as amplifiers, FETs are usually biased in the saturation region.
作为放大器，场效应管通常工作于饱和区。

3 The amplifier gain remains almost constant over the midfrequency band. It falls off at high frequencies where transistor-model capacitances no longer have very high reactances. For AC amplifiers the gain falls off at low frequencies as well because the coupling and bypass capacitances no longer have very low reactances.
放大电路的放大倍数在中频段几乎保持不变，在高频段，由于晶体管的电容容抗降低，放大倍数会下降，在低频段，由于耦合电容容抗升高，放大倍数也会下降。

第六节 模拟集成电路
Section 6　Analog Integrated Circuits

保护电路　protecting circuit
比例电流源　scaling current source
差分放大电路　differential amplifier
差模放大倍数　differential-mode gain
差模信号　differential-mode signal
单极型集成电路　unipolar integrated circuit
等效电路　equivalent circuit
多路电流源　multiple current source
共模放大倍数　common-mode gain
共模信号　common-mode signal
共模抑制比　common-mode rejection ratio
恒流源　constant current source
集成电路特点　feature of integrated circuit
集成运算放大电路　integrated operational amplifier circuit
截止频率　3-dB frequency
镜像电流源　current mirror
开环增益　open-loop gain
理想运放　ideal operational amplifier
失调电流　offset current
失调电压　offset voltage
通用型　popular type
同相输入端　non-inverting input terminal
威尔逊电流源　Wilson current source

性能扩展	performance extension	运放	operational amplifier
有源负载	active load	专用型	special type

All signals can be considered to be composed of sine waves of various amplitudes, frequencies, and Phases. To amplify a signal without distortion, the amplifier must have constant gain for all of the frequencies contained in the signal.

所有的信号都可以看成是将各种振幅、频率和相位的正弦波组成。放大信号不失真,放大器必须对所有频率的信号有恒定增益。

第七节 放大电路中的反馈
Section 7　Feedback in Amplifier

闭环增益	closed-loop gain	干扰	interference
电流并联	current-parallel	共模抑制比	common-mode rejection ratio
电流串联	current-series	环路增益	loop gain
电压并联	voltage-parallel	基本方程	basic equation
电压串联	voltage-series	深度负反馈的特点	property of circuit with strong negative feedback
反馈放大电路	feedback amplifier	深度负反馈电路	circuit with strong negative feedback
反馈概念	feedback concept		
反馈极性判断	feedback polarity examination	稳定判据	stability criterion
反馈通路	feedback path	校正措施	compensation method
反馈网络	feedback network	引入负反馈的一般原则	general rules for introducing negative feedback
反馈信号	feedback signal		
反馈组态	feedback configuration	正反馈	positive feedback
分析方法	method of analysis	自激振荡	self-excited oscillation
负反馈	negative feedback	自举电路	bootstrapping circuit
负反馈电路的自激振荡	self-excited oscillation of feedback circuit		

Negative feedback is employed to make the amplifier gain less sensitive to component variations, In order to control input and output impedances, to extend bandwidth, to reduce nonlinear distortion, and to enhance signal-to-noise (and signal-to-interference) ratio.

负反馈放大电路能够稳定放大倍数、改变输入和输出电阻、扩展频带、减小非线性失真和提高信噪比。

第八节 功率放大电路
Section 8　Power Amplifier Circuit

BTL 电路	balanced transformerless circuit	集成功放	integrated circuit power amplifier
倍压整流电路	voltage doubler circuit	甲类放大	class A amplification
丙类	class C	交越失真	crossover distortion
达林顿复合管	Darlington connection	散热	heat dissipation
低频功率放大电路	low frequency power amplifier	推挽	push-pull
功率放大电路	power amplifier	最大输出功率	maximum output power
互补电路	complementary pair		

Output stages are classified according to the transistor conduction angle: class A (360°), class AB (slightly greater than 180°), class B(180°), and class C(smaller than 180°).

功率放大电路的输出级根据晶体管的导通角分类:甲类(导通角 360°)、甲乙类(导通角稍大于 180°)、乙类(导通角 180°)和丙类(导通角小于 180°)。

第九节 信号的处理与信号的产生电路
Section 9 Signal Processing and Signal Generation Circuit

T 型反馈网络　T type feedback network
巴特沃思滤波器　Butterworth filter
倍频　frequency multiplication
倍频器　frequency multiplier
比例运算电路　scaling circuit
变压器反馈式　transformer feedback
变压器反馈式正弦波振荡电路　transformer feedback sinusoidal oscillator
波形变换电路　waveform converter
波形发生电路　waveform generator
差分输入　differential input
乘法运算电路　multiplication circuit
除法运算电路　division circuit
串并联（文氏桥）式振荡器　Wien bridge oscillator
带通滤波器　band-pass filter
带阻滤波器　band-reject filter
低通滤波器　low-pass filter
电感反馈式（电感三点式）　Hartley feedback
电感反馈式正弦波振荡电路　Hartley feedback sinusoidal oscillator
电容反馈式（电容三点式）　Colpitts feedback
电压比较器　voltage comparator
电压跟随器　voltage follower
对数运算电路　logarithmic circuit
反相输入　inverting input
方波发生器　square wave generator
分频　frequency division
分频器　frequency divider
幅值　amplitude
高通滤波器　high-pass filter
函数发生器　function generator
混频器　mixer
积分运算电路　integration circuit
集成电压比较器　integrated voltage comparator
简单比较器　simple comparator
晶片　wafer
矩形波发生器　rectangular wave generator
锯齿波发生器　sawtooth wave generator
品质因数　quality factor
求和电路　summing circuit
三角波变锯齿波　triangular wave-sinusoidal wave converter
三角波发生电路　triangular wave generator
石英晶体　crystal oscillator
瞬态响应　transient response
特征频率　characteristic frequency
通带电压放大倍数　pass-band voltage gain
通带截止频率　cut-off frequency of pass-band
同相输入　noninverting input
外特性　external characteristic
微分电路　differential circuit
限幅　clipping
限幅电路　clipping circuit
信号转换电路　signal transfer circuit
选频　frequency-selection
选频网络　frequency-selective network
压控电压源　voltage-controlled voltage source
阈值电压　threshold voltage
运算电路　operational circuit
占空比　duty ratio
振荡条件　condition of oscillation
正弦波振荡电路　sinusoidal oscillator circuit
指数电路　exponential circuit
滞回比较器　regenerative comparator
中心频率　center frequency
状态变量　state variable

1 Filters perform a frequency-selection function: passing signals whose frequency spectrum lies within a specified range, and stopping signals whose frequency spectrum falls outside this range.
滤波电路具有选频的功能：通过通带内的信号，滤除掉阻带的信号。

2 The basic structure of a sinusoidal oscillator consists of an amplifier and a frequency-selective network connected in a positive-feedback loop.
正弦波振荡电路由放大电路、选频网络和正反馈电路组成。

第十节 直流电路
Section 10 DC Circuit

安全区保护 safety operating area protection	基准源 reference source
半波整流 half wave rectifier	开关型直流电源 switching mode direct power supply
并联型 parallel type	
串联型 serial type	滤波电容 filtering capacitance
串联型稳压器 series voltage regulator	脉动系数 impulsing factor
单片集成稳压器 monolithic integrated regulator	桥式整流电路 bridge rectifier circuit
电容滤波 capacitance filter	外特性 external characteristic
过流保护 current overload protection	稳压电源 regulated power supply
过热保护 thermal overload protection	稳压管稳压器 zener voltage regulator

1 The regulated power supply consists of transformer, rectifier circuit, filter circuit and voltage regulator.

稳压电源由变压器、整流电路、滤波电路和稳压器组成。

2 Rectifier circuits are useful in battery-charging circuits and for converting AC into nearly constant DC to power electronic circuits such as amplifier.

整流电路通常用于电池充电和把交流电转换成几乎恒定不变的直流电的电路中,如放大器中的电子电源。

第二部分　应用篇

第二十三章　应用范例
Chapter 23　Application Examples

第一节　摘要写作
Section 1　Abstract Writing

1. What is an abstract?

An abstract is an independent statement that briefly conveys the essential information of a paper; presents the objective, methods, results, and conclusions of a research project.

什么是研究摘要？

摘要是一个独立性的描述，简洁地传达了论文的重要信息，提出了研究项目的目的、方法、结果和结论。

2. Parts of Abstract　摘要的组成

(1) background information　背景信息

One or two simple opening sentences can be used to make clear the working context.

一或两个开场句子交代工作背景。

(2) purpose of the study / principal activity　研究目的或主要研究活动

One or two sentences can be used to provide the purpose of the work.

一或两个句子给出工作目的。

(3) method　研究方法

One or two sentences can be used to explain what problems have already been solved.

一或两个句子阐述已经解决的问题。

(4) research results　研究结果

One or two sentences can be used to indicate the main findings.

一或两个句子表明主要的结果。

(5) conclusion / recommendation　结论或建议

Use one sentence to state clearly the most important conclusion of the work.

一或两个句子表明研究工作的最重要的结论。

3. Sample Analysis　实例分析

Abstract: With a listening typewriter, what an author says would be automatically recognized and displayed in front of him or her. However, speech recognition is not yet advanced enough to provide people with a reliable listening typewriter. An aim of our experiments was to determine if an imperfect listening typewriter would be used for composing letters. Participants dictated letters, either in isolated words or in consecutive word speech. They did this with simulations of listening typewriters that recognized either a limited vocabulary or an unlimited vocabulary. Results indicated that some versions, even upon first using, were at least as good as traditional methods of handwriting and dictating. Isolated word speech with large vocabulary may provide the basis for useful listening typewriter.

摘要：使用收听打字机可以自动识别作者说的话，并且显示在作者面前。但是，话语识别技术尚未达到能为人们提供可靠的收听打字机的程度。我们做此试验的一个目的就是要确定是否能用技术上尚不完善的收听打字机来写信。参试人员口授单个词语或连续的话语。他们既模拟试验能识别有限制的词语的收听打字机,也试验能识别无限制词语的收听打字机。结果表明,有些完成的版本,即使是首次使用,也至少和传统的手写和口授方法一样好。大词量的不连贯词语构成的话语可以为实用收听打字机奠定基础。

(1) With a listening typewriter, what an author says would be automatically recognized and displayed in front of him or her. （General background 一般背景）

(2) However, speech recognition is not yet advanced enough to provide people with a reliable listening typewriter. (Specific background 特定背景)

(3) An aim of our experiments was to determine if an imperfect listening typewriter would be used for composing letters. (Purpose 研究目的)

(4) Participants dictated letters, either in isolated words or in consecutive word speech. （Methods 研究方法)

(5) They did this with simulations of listening typewriters that recognized either a limited vocabulary or an unlimited vocabulary. (Methods 研究方法)

(6) Results indicated that some versions, even upon first using, were at least as good as traditional methods of handwriting and dictating. (Research Result 研究结果)

(7) Isolated word speech with large vocabulary may provide the basis for useful listening typewriter. (Conclusion 结论)

4. Verb Tense for Abstracts 摘要的动词时态

(1) background information: present tense

背景信息：一般现在时

Example: In an increasing number of second language (L2) classrooms, teachers and researchers are taking on new roles and responsibilities.

例子：在越来越多的第二外语课堂,教师和研究人员正在承担着新的角色和责任。

(2) purpose of the study / principal activity: past, present perfect, past future, past future perfect

研究目的或主要研究活动：一般过去时,过去完成时、过去将来时或过去将来完成时

Example: An aim of our experiments was to determine if an imperfect listening typewriter would be useful for composing letters.

例子：我们做此试验的一个目的就是要确定是否能用技术上尚不完善的收听打字机来写信。

(3) methods: past tense, passive voice is often used

研究方法：一般过去时,经常使用被动语态

Example: Most of the activities during the day were tape-recorded. And the data were supplemented with interviews.

例子：今天的大部分活动用磁带记录下来了,而且采访的数据也补充了。

(4) research results: past tense

研究结果：一般过去时

Example: Results indicated that some versions, even upon first using them, were at least as good as traditional methods of handwriting and dictating.

例子：结果表明,有些完成的版本,即使是首次使用,也至少和传统的手写和口授方法一样好。

(5) Conclusions: present tense

结论：一般现在时

Example: Isolated word speech with large vocabulary may provide the basis for a useful listening typewriter.

例子：大词量的不连贯词语构成的话语可以为实用收听打字机奠定基础。

第二节 实验报告
Section 2　Experimental Report

（课程名称：电路原理实验）
(Course name：Experiments on Principle of Electric Circuit)

学院：机电信息工程学院
College：College of Mechanical and Electrical Information Engineering
系：自动化
Department：Automation
专业：自动化
Major：Automation
学生姓名：于伟
Student's Name：Yu Wei
班级：自动化 101
Class：Automation 101
学号：2010 023 127
Student ID Number：2010 023 127
实验地点：综 C-705
Experiment Site：Complex Building C-705
组号：1
Group Number：1
实验性质：验证性
Experimental Nature：Confirmatory
实验时间：2011 年 10 月 16
Time：October 16th, 2011

Experiment：　Kirchhoff's Law and Superposition Theorem
实验名称：基尔霍夫定律和叠加原理

1. Experiment purpose
To get a better understanding of the Kirchhoff's Law and the content and the scope of superposition theorem.

1. 实验目的
加深对基尔霍夫定律和叠加原理的内容和适用范围的理解。

2. Laboratory instruments
Electric experimental devices：DG013T, DY031T, and DG054-1T

2. 实验仪器
电工实验装置：DG013T、DY031T、DG054-1T

3. Experimental theory
3.1　Kirchhoff's Law is the basic law of lumped circuit. It includes circuit law and voltage law.
Kirchhoff's current Law(KCL) states that in the lumped circuit, at any time, at any node, the algebraic sum of currents entering a node (or a closed boundary) is zero, namely：
$$\Sigma I = 0$$
Kirchhoff's Voltage Law (KVL) states that in the lumped circuit, at any time, the algebraic sum of all voltages around any closed path (or loop) in a circuit equals zero, namely：
$$\Sigma U = 0$$
3.2　Superposition Theorem is an important theorem of linear circuit.

The superposition principle states that the voltage across (or current through) an element in a linear circuit is the algebraic sum of the voltages across (or currents through) that element due to each independent source acting alone.

3. 实验原理

3.1 基尔霍夫定律是集总电路的基本定律。它包括电流定律和电压定律。

基尔霍夫电流定律:在集总电路中,任何时刻,流入结点(或闭合面)的电流的代数和恒等于零,即:

$$\sum I = 0$$

基尔霍夫电压定律:在集总电路中,任何时刻,沿任一闭合路径(或回路)内所有电压的代数和恒等于零,即:

$$\sum U = 0$$

3.2 叠加原理是线性电路的一个重要定理。

叠加原理:线性电路中,任一器件上的电压或电流都是电路中各个独立电源单独作用时,在该处产生的电压或电流的代数和。

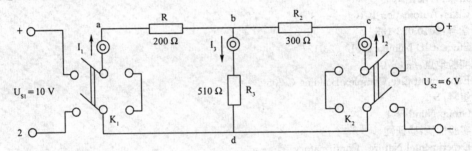

Fig. 1 Experimental Circuit
图 1 实验线路

4. Experimental procedures

4.1 Kirchhoff's Law

4.1.1 Connect the circuit as shown in Fig. 1. I_1, I_2, I_3 are circuit sockets. K_1、K_2 are double-pole-double-throw switches.

4.1.2 Turn K_1、K_2 to short circuit side, and regulate the regulated power supply, make U_{S1} = 10V, U_{S2} = 6V, (use 20V DC voltmeter of DG054-1T to measure the voltage output of DY031T separately).

4.1.3 Turn K_1、K_2 to power supply side, measure and take notes with the parameter in Table 1 and Table 2. Verify Kirchhoff's Law.

Table 1 Kirchhoff Circuit Law

I_1 (mA)	I_2 (mA)	I_3 (mA)	Verify node b: $\sum I = 0$
15.7	−2.5	−13.2	$\sum I = 0$

Table 2 Kirchhoff Voltage Law

U_{ab}(V)	U_{bc}(V)	U_{bd}(V)	U_{da}(V)	U_{cd}(V)	Verify $\sum U = 0$	
					Loop abcda	Loop abda
3.18	0.81	6.82	−10.00	6.01	$\sum U = 0$	$\sum U = 0$

4.2 Superposition Theorem

The experimental circuit is shown in Fig. 1.

4.2.1 Turn K_2 to short circuit side, K_1 to power supply side, and make U_{S1} function alone. The voltages across and currents through each of the elements are then measured and recorded in Table 3.

4.2.2 Turn K_1 to short circuit side, K_2 to power supply side, let the circuit be energized by U_{S2} only, and then measure the circuits through and the voltages across each of the elements, and put them into Table 3.

4.2.3 The data of the two power supply functioning together can be acquired from procedure 1.

Table 3　Superposition Theorem

	I_1(mA)	I_2(mA)	I_3(mA)	U_{ab}(V)	U_{bc}(V)	U_{bd}(V)
U_{S1} Functions alone	25.3	−16.1	−9.4	5.08	4.89	4.89
U_{S2} Functions alone	−9.7	13.4	−3.7	−1.96	−4.03	1.97
U_{S1}, U_{S2} Interaction	15.7	−2.5	−13.2	3.17	0.81	6.82
Verify Superposition Principle	25.3+(−9.7)= 15.6	−16.1+13.4= −2.7	−9.4+(−3.7)= −13.1	3.12	0.86	6.86

4. 实验步骤

4.1 基尔霍夫定律

4.1.1　按图 1 接线。其中 I_1、I_2、I_3 是电流插口，K_1、K_2 是双刀双掷开关。

4.1.2　先将 K_1、K_2 合向短路线一边，调节稳压电源，使 $U_{S1}=10V$，$U_{S2}=6V$，(用 DG054-1T 的 20V 直流电压表来分别测量 DY031T 的输出电压)。

4.1.3　将 K_1、K_2 合向电源一边，按表 1 和表 2 中给出的各参量进行测量并记录，验证基尔霍夫定律。

表 1　基尔霍夫电流定律

I_1(mA)	I_2(mA)	I_3(mA)	验证节点 b: $\sum I=0$
15.7	−2.5	−13.2	$\sum I=0$

表 2　基尔霍夫电压定律

U_{ab}(V)	U_{bc}(V)	U_{bd}(V)	U_{da}(V)	U_{cd}(V)	验证 $\Sigma U=0$	
					回路 abcda	回路 abda
3.18	0.81	6.82	−10.00	6.01	$\sum U=0$	$\sum U=0$

4.2 叠加原理

实验电路如图 1 所示。

4.2.1　把 K_2 掷向短路线一边，K_1 掷向电源一边，使 U_{S1} 单独作用，测量各电流、电压并记录于表 3 中。

4.2.2　把 K_1 掷向短路线一边，K_2 掷向电源一边，使 U_{S2} 单独作用，测量各电流、电压并记录在表 3 中。

4.2.3　两电源共同作用时的数据在实验步骤 1 中获取。

表 3　叠加原理

	I_1(mA)	I_2(mA)	I_3(mA)	U_{ab}(V)	U_{bc}(V)	U_{bd}(V)
U_{S1} 单独作用	25.3	−16.1	−9.4	5.08	4.89	4.89
U_{S2} 单独作用	−9.7	13.4	−3.7	−1.96	−4.03	1.97
U_{S1}、U_{S2} 共同作用	15.7	−2.5	−13.2	3.17	0.81	6.82
验证叠加原理	25.3+(−9.7)=15.6	−16.1+13.4=−2.7	−9.4+(−3.7)=−13.1	3.12	0.86	6.86

5. Data analysis and conclusion

In table 1, we can conclude from the experimental data obtained that node b met the equation $\sum I = I_1 + I_2 + I_3 = 0$. The experimental KCL results verified KCL. That is, in the lumped circuit, at any time, the algebraic sum of all branches currents entering a node equals to zero constantly.

In table 2, for loop abcda, the equation $U_{ab} + U_{bc} + U_{cd} + U_{da} = 3.18 + 0.81 + 6.01 + (-10.00) = 0$ verified that Kirchhoff's voltage law was valid. That is, in the lumped circuit, at any time, the algebraic sum of all the voltages around any closed path (loop or mesh) is zero.

The experimental superposition principle results shown in table 3 did not meet the predicted values. All the equations produced values within 0.5 volt(or ampere) away from the measured values, but the errors were still small enough to conclude that superposition principle is valid.

5. 数据分析及结论

表1,由实验所得数据可得:对于结点 b 满足 $\sum I = I_1 + I_2 + I_3 = 0$,验证基尔霍夫电流定律成立。即在集总参数电路中,任何时刻,对任一结点,所有支路电流的代数和恒等于零。

表2,对于回路 abcda 有 $U_{ab} + U_{bc} + U_{cd} + U_{da} = 3.18 + 0.81 + 6.01 + (-10.00) = 0$。验证基尔霍夫电压定律成立。即在集总电路中,任何时刻,沿任一闭合路径(回路或网孔)内所有电压的代数和恒等于零。

表3,由实验数据可知,虽然测量值并不满足预期值,计算结果和测量值有 0.5 伏(或安培)以内的误差,但此误差在允许范围内,所以叠加原理成立。

第二十四章 设备使用手册
Chapter 24　Equipment Manual

数字存储示波器
Digital Storage Oscilloscope

Operating Basics
操作基础
Display Area
显示区域

In addition to displaying waveforms, the display is filled with many details about the waveform and the oscilloscope control settings.

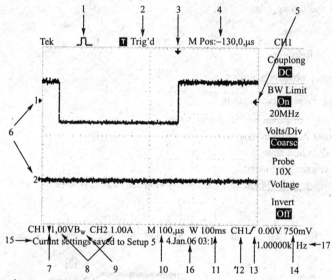

1. Icon display shows acquisition mode.

　⊓ Sample mode

　⊓ Peak detect mode

　⊓ Average mode

2. Trigger status indicates the following:

　[Armed] The oscilloscope is acquiring pretrigger data. All triggers are ignored in this state.

　[Ready] All pretrigger data has been acquired and the oscilloscope is ready to accept a trigger.

　[Trig'd] The oscilloscope has seen a trigger and is acquiring the posttrigger data.

　[Stop] The oscilloscope has stopped acquiring waveform data. .

　[Auto] The oscilloscope is in auto mode and is acquiring waveforms in the absence of triggers.

　[Scan] The oscilloscope is acquiring and displaying waveform data continuously in scan mode.

3. Marker shows horizontal trigger position. Turn the Horizontal Position knob to adjust the position of the marker.

4. Readout shows the time at the center graticule. The trigger time is zero.

5. Marker shows edge or pulse width trigger level.

6. On-screen markers show the ground reference points of the displayed waveforms. If there is no marker, the channel is not displayed.

7. An arrow icon indicates that the waveform is inverted.

8. Readouts show the vertical scale factors of the channels.

9. A BW icon indicates that the channel is bandwidth limited.

10. Readout shows main time base setting.

11. Readout shows window time base setting if it is in use.

12. Readout shows trigger source used for triggering.

13. Icon shows selected trigger type as follows:

⌐ Edge trigger for the rising edge

⌐ Edge trigger for the falling edge

⌐ Video trigger for line sync

⌐ Video trigger for field sync

⌐ Pulse width trigger, positive polarity

⌐ Pulse width trigger, negative polarity

14. Readout shows edge or pulse width trigger level.

15. Display area shows helpful messages.

16. Readout shows date and time.

17. Readout shows trigger frequency.

除显示波形外,显示屏上还含有很多关于波形和示波器控制设置的详细信息

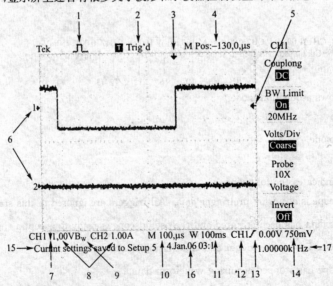

1. 显示图标表示采集模式。

⌐ 取样模式

第二十四章 设备使用手册

⎍ 峰值检测模式

⎍ 均值模式

2.触发状态显示如下：

▯Armed 示波器正在采集预触发数据。在此状态下忽略所有触发。

▯Ready 示波器已采集所有预触发数据并准备接受触发。

▯Trig'd 示波器已被触发，正在采集触发后的数据。

▯Stop 示波器已停止采集波形数据。

▯Auto 示波器处于自动模式并在无触发状态下采集波形。

▯Scan 在扫描模式下示波器连续采集并显示波形。

3. 使用标记显示水平触发位置。旋转"水平位置"旋钮调整标记位置。
4. 用读数显示中心刻度线的时间。触发时间为零。
5. 使用标记显示"边沿"脉冲宽度触发电平。
6. 使用屏幕标记表显示波形的接地参考点。如没有标记，不会显示通道。
7. 箭头图标表示波形是反相的。
8. 以读数显示通道的垂直刻度系数。
9. BW 图标表示通道是带宽限制的。
10. 以读数显示主时基设置。
11. 如使用窗口时基，以读数显示窗口时基设置。
12. 以读数显示触发使用的触发源。
13. 采用图标显示以下选定的触发类型：

⎍ 上升沿的"边沿"触发。

⎍ 下降沿的"边沿"触发。

⎍ 行同步的"视频"触发。

⎍ 场同步的"视频"触发。

⎍ "脉冲宽度"触发，正极性。

⎍ "脉冲宽度"触发，负极性。

14. 用读数表示边沿或脉冲宽度触发电平。
15. 显示区显示有用信息。
16. 以读数显示时间和日期。
17. 以读数显示触发频率。

Understanding Oscilloscope Functions

To use your oscilloscope effectively, you need to learn about the following functions:

(1) Setting up the oscilloscope
(2) Triggering
(3) Acquiring signals (waveforms)
(4) Scaling and positioning waveforms
(5) Measuring waveforms

(1) Setting Up the Oscilloscope

You should become familiar with several functions that you may often use when operating your oscilloscope: using autoset, saving a setup, and recalling a setup, etc.

Using Autoset Each time you push the Autoset button, the Autoset function obtains a stable waveform display for you. It automatically adjusts the vertical scale, horizontal scale and trigger settings. Autoset also displays several automatic measurements in the graticule area, depending on the signal type.

Saving a Setup The oscilloscope saves the current setup if you wait five seconds after the last change before you power off the oscilloscope. The oscilloscope recalls this setup the next time you apply power. You can use the Save/Recall Menu to save up to ten different setups.

Recalling a Setup The oscilloscope can recall the last setup before the oscilloscope was powered off, any saved setups, or the default setup.

Default Setup The oscilloscope is set up for normal operation when it is shipped from the factory. This is the default setup. To recall this setup, one has to push the Default Setup button.

(2) Triggering

The trigger determines when the oscilloscope starts to acquire data and to display a waveform. When a trigger is set up properly, the oscilloscope converts unstable displays or blank screens into meaningful waveforms.

When you push the Run/Stop or Single button to start an acquisition, the oscilloscope goes through the following steps:

1. Acquires enough data to fill the portion of the waveform record to the left of the trigger point. This is called the pretrigger.
2. Continues to acquire data while waiting for the trigger condition to occur.
3. Detects the trigger condition.
4. Continues to acquire data until the waveform record is full.
5. Displays the newly-acquired waveform.

Source

You can use the Trigger Source options to select the signal that the oscilloscope uses as a trigger. The source can be the AC power line (available only with Edge triggers), or any signal connected to a channel BNC or to the Ext Trig BNC.

Types

The oscilloscope provides three types of triggers: Edge, Video, and Pulse Width.

Modes

You can select the Auto or the Normal trigger mode to define how the oscilloscope acquires data when it does not detect a trigger condition.

To perform a single sequence acquisition, one has to push the Single button.

Coupling

You can use the Trigger Coupling option to determine which part of the signal will pass to the trigger circuit. This can help you attain a stable display of the waveform.

To use trigger coupling, push the Trig Menu button, select an Edge or Pulse trigger, and select a Coupling option.

Position

The horizontal position control establishes the time between the trigger and the screen center. Refer to *Horizontal Scale and Position*, *Pretrigger Information* for information on how to use this control to position the trigger.

Slope and Level

The Slope and Level controls help to define the trigger. The Slope option (Edge trigger type only) determines whether the oscilloscope finds the trigger point on the rising or the falling edge of a signal. The Trigger Level knob controls where on the edge the trigger point occurs.

(3) Acquiring Signals

When you acquire a signal, the oscilloscope converts it into a digital form and displays a waveform. The acquisition mode defines how the signal is digitized, and the time base setting affects the time span and level of detail in the acquisition.

Acquisition Modes

There are three acquisition modes: Sample, Peak Detect, and Average.

Sample

In this acquisition mode, the oscilloscope samples the signal in evenly spaced intervals to construct the waveform. This mode accurately represents signals most of the time. However, this mode does not acquire rapid variations in the signal that may occur between samples. This can result in aliasing, and may cause narrow pulses to be missed. In these cases, you should use the Peak Detect mode to acquire data.

Peak Detect

In this acquisition mode, the oscilloscope finds the highest and lowest values of the input signal over each sample interval and uses these values to display the waveform. In this way, the oscilloscope can acquire and display narrow pulses, which may have otherwise been missed in Sample mode. Noise will appear to be higher in this mode.

Average

In this acquisition mode, the oscilloscope acquires several waveforms, averages them, and displays the resulting waveform. You can use this mode to reduce random noise.

Time Base

The oscilloscope digitizes waveforms by acquiring the value of an input signal at discrete points. The time base allows you to control how often the values are digitized. To adjust the time base to a horizontal scale that suits your purpose, use the horizontal scale knob.

(4) Scaling and Positioning Waveforms

You can change the display of waveforms by adjusting the scale and position. When you change the scale, the waveform display will increase or decrease in size. When you change the position, the waveform will move up, down, right, or left. The channel indicator (located on the left of the graticule) identifies each waveform on the display. The indicator points to the ground reference level of the waveform record. You can view the display area and readouts.

Vertical Scale and Position

You can change the vertical position of waveforms by moving them up or down in the display. To compare data, you can align a waveform above another or you can align waveforms on top of each other.

You can change the vertical scale of a waveform. The waveform display will contract or expand relative to the ground reference level.

Horizontal Scale and Position; Pretrigger Information

You can adjust the Horizontal Position control to view waveform data before the trigger, after the trigger, or some of each. When you change the horizontal position of a waveform, you are actually changing the time between the trigger and the center of the display. (This appears to move the waveform to the right or left on the display.)

Time Domain Aliasing

Aliasing occurs when the oscilloscope does not sample the signal fast enough to construct an accurate waveform record. When this happens, the oscilloscope displays a waveform with a frequency lower than the actual input waveform, or triggers and displays an unstable waveform.

The oscilloscope accurately represents signals, but is limited by the probe bandwidth, the oscilloscope bandwidth, and the sample rate. To avoid aliasing, the oscilloscope must sample the signal more than twice as fast as the highest frequency component of the signal.

The highest frequency that the oscilloscope sampling rate can theoretically represent is the Nyquist frequency. The sample rate is called the Nyquist rate, and is twice the Nyquist frequency.

There are several ways to check for aliasing:

Turn the horizontal scale knob to change the horizontal scale. If the shape of the waveform changes drastically, you may have aliasing.

Select the Peak Detect acquisition mode. This mode samples the highest and lowest values so that the oscilloscope can detect faster signals. If the shape of the waveform changes drastically, you may have aliasing.

If the trigger frequency is faster than the display information, you may have aliasing or a waveform that crosses the trigger level multiple times. Examining the waveform allows you to identify whether the shape of the signal is going to allow a single trigger crossing per cycle at the selected trigger level.

If multiple triggers are likely to occur, select a trigger level that will generate only a single trigger per cycle. If the trigger frequency is still faster than the display indicates, you may have aliasing.

If the trigger frequency is slower, this test is not useful.

If the signal you are viewing is also the trigger source, use the graticule or the cursors to estimate the frequency of the displayed waveform. Compare this to the Trigger Frequency readout in the lower right corner of the screen. If they differ by a large amount, you may have aliasing.

(5) Taking Measurements

The oscilloscope displays graphs of voltage versus time and can help you to measure the displayed waveform. There are several ways to take measurements. You can use the graticule, the cursors, or an automated measurement.

Graticule This method allows you to make a quick, visual estimate. For example, you might look at a waveform amplitude and determine that it is a little more than 100 mV.

You can take simple measurements by counting the major and minor graticule divisions involved and multiplying by the scale factor.

For example, if you counted five major vertical graticule divisions between the minimum and maximum values of a waveform and knew you had a scale factor of 100 mV/division, then you could calculate your peak-to-peak voltage as follows:

5 divisions $\times$ 100 mV/division $=$ 500 mV

Cursors This method allows you to take measurements by moving the cursors, which always appear in pairs, and reading their numeric values from the display readouts. There are two types of cursors: Amplitude and Time. When you use cursors, be sure to set the Source to the waveform on the display that you want to measure. To use cursors, one has to push the Cursor button.

Amplitude Cursors Amplitude cursors appear as horizontal lines on the display and measure the vertical parameters. Amplitudes are referenced to the reference level. For the Math FFT function, these cursors measure magnitude.

Time Cursors Time cursors appear as vertical lines on the display and measure both horizontal and vertical parameters. Times are referenced to the trigger point. For the Math FFT function, these cursors measure frequency.

Time cursors also include a readout of the waveform amplitude at the point the waveform crosses the cursor.

Automatic The Measure Menu can take up to five automatic measurements. When you take automatic measurements, the oscilloscope does all the calculating for you. Because the measurements use the waveform record points, they are more accurate than the graticule or cursor measurements. Automatic measurements use readouts to show measurement results. These readouts are updated periodically as the oscilloscope acquires new data.

了解示波器的功能
为了有效地使用示波器,需要了解示波器的以下功能:
(1)设置示波器
(2)触发
(3)采集信号(波形)
(4)缩放并定位波形
(5)测量波形

(1)设置示波器
操作示波器时,应熟悉一些经常使用的功能:自动设置、储存设置和调出设置。

自动设置
使用"自动设置"功能可获得稳定的波形显示效果。它可以自动调整垂直刻度、水平刻度和触发设置。自动设置也可在刻度区域显示几个自动测量结果,这取决于信号类型。

储存设置
关闭示波器电源前,如果在最后一次更改后已等待五秒钟,示波器就会存储当前设置。下次接通电源时,示波器会调出此设置。

调出设置
示波器可以调出关闭电源前的最后一个设置、存储的任何一个设置或默认设置。

默认设置
示波器在出厂前被设置为用于常规操作,这就是默认设置。要想调出该设置,需按下"默认设置"键。

(2)触发
触发器将确定示波器开始采集数据和显示波形的时间。正确设置触发器后,示波器就能将不稳定的显示结果或空白显示屏转换为有意义的波形。
当按下"运行、停止"或"单次序列"按钮开始采集时,示波器执行下列步骤:
1. 采集足够的数据来填充触发点左侧的波形记录部分,这也被称为预触发。
2. 在等待触发条件出现的同时继续采集数据。
3. 检测触发条件。
4. 在波形记录填满之前继续采集数据。
5. 显示最近采集的波形。

信源
可以使用"触发源"选项来选择示波器用作触发源的信号。信源可以是连接到交流信号线(仅用于边沿触发),或者连接到通道 BNC 或者外部触发 BNC 的其他信号。

类型
示波器提供三类触发:边沿、视频和脉冲宽度。

模式
在示波器未检测到触发条件时,可以选择一种"触发模式"来定义示波器采集数据的方式。模式有"自动"和"正常"两种。

耦合
可使用"触发耦合"选项确定哪一部分信号将通过触发电路。这有助于获得一个稳定的显示波形。要使用触发耦合,可按下"触发菜单"按钮,选择一个"边沿"或"脉冲"触发,然后选择一个"耦合"选项。

位置
水平位置控制可确定触发位置与显示屏中心之间的时间。
参考"水平比例和位置;预触发信息"以获取如何应用此控制使触发器就位的相关信息。

斜率和电平
"斜率"和"电平"控制有助于定义触发器。"斜率"选项(仅限于"边沿"触发类型)确定示波器是在信号的上升边沿还是在下降边沿找到触发点。"触发电平"旋钮控制触发点在边沿的什么位置上出现。

(3) 采集信号

采集信号时,示波器将其转换为数字形式并显示波形。采集模式定义采集过程中信号被数字化的方式和时基设置影响采集的时间跨度和细节程度。

采集模式

有三种采集模式:取样、峰值检测和平均取样。

取样

在这种采集模式下,示波器以均匀时间间隔对信号进行取样以建立波形。此模式多数情况下可以精确表示信号。然而,此模式不能采集取样之间可能发生的快速信号变化。这可以导致假波现象并可能漏掉窄脉冲。在这些情况下,应使用"峰值检测"模式来采集数据。

峰值检测

在这种采集模式下,示波器在每个取样间隔中找到输入信号的最大值和最小值,并使用这些值显示波形。这样,示波器就可以采集并显示窄脉冲,否则这些窄脉冲在"取样"模式下可能已被漏掉。在这种模式下,噪声看起来似乎更大。

平均取样

在这种采集模式下,示波器采集几个波形,将它们平均,然后显示最终波形。可以使用此模式来减少随机噪声。

时基

示波器通过在不连续点处采集输入信号的值来数字化波形,使用时基可以控制这些数值被数字化的频度。要将时基调整到某一水平刻度以适应您的要求,可使用"秒/格"旋钮。

(4) 缩放并定位波形

可以调整波形的比例和位置来更改显示的波形。改变比例时,波形显示的尺寸会增加或减小。改变位置时,波形会向上、向下、向右或向左移动。通道指示器(位于刻度的左侧)会标识显示屏上的每个波形。指示器指向所记录波形的接地参考电平。显示器有显示区域和读数。

垂直刻度和位置

通过在显示屏上向上或向下移动波形来更改其垂直位置。要比较数据,可以将一个波形排列在另一个波形的上面,或者可以把波形相互叠放在一起。

可以更改某个波形的垂直比例。显示的波形将基于接地参考电平进行缩放或扩大。

水平刻度和位置;预触发信息

可以调整"水平位置"控制来查看触发前、触发后或触发前后的波形数据。改变波形的水平位置时,实际上改变的是触发位置和显示屏中心之间的时间。(这看起来就像在显示屏上向右或向左移动波形。)

时域假波现象　如果示波器对信号进行采样时不够快,从而无法建立精确的波形记录时,就会有假波现象。此现象发生时,示波器将以低于实际输入波形的频率显示波形,或者触发并显示不稳定的波形。

示波器精确表示信号的能力受到探头带宽、示波器带宽和采样速率的限制。要避免假波现象,示波器的采样频率必须至少比信号中的最高频率分量快两倍。

示波器采样速率在理论上所能表示的最高频率就是奈奎斯特频率。采样速率被称为奈奎斯特速率,是奈奎斯特频率的两倍。

有多种方法可检查假波现象:

旋转"秒/格"旋钮可以改变水平比例。如果波形形状剧烈变化,则可能有假波现象。

选择"峰值检测"获取方式。在此方式下,将对最大值和最小值进行采样,因此示波器可以检测速度更快的信号。如果波形形状剧烈变化,则可能有假波现象。

如果触发频率比显示信息的速度快,就可能有假波现象或出现波形多次跨过触发电平的情况。通过检查波形,可能会发现在选定的触发级上,信号的形状是否允许在每个周期内触发一次。

如果可能发生多次触发,则选择某一触发电平,使每个周期仅发生一次触发。如果触发频率仍比显示速度快,就可能有假波现象。

如果触发频率比较慢,这种测试就不起作用。

如果正观察的信号也是触发源,则使用刻度或光标来估计所显示波形的频率,并与显示屏右下角的

"触发频率"读数相比较。如果它们相差很大,则可能有假波现象。

(5)测量波形

示波器将显示电压相对于时间的图形并帮助您测量显示波形。有几种测量方法。可以使用刻度、光标进行测量或执行自动测量。

刻度

使用此方法能快速、直观地做出估计。例如,可以观察波形幅度,确定它是否略高于100mV。

可通过计算相关的大、小刻度分度并乘以比例系数来进行简单的测量。

例如,如果计算出在波形的最大值和最小值之间有五个主垂直刻度分度,并且已知比例系数为100毫伏/格,则可按照下列方法来计算峰峰值电压:5 格×100 毫伏/格＝ 500 毫伏。

光标

使用此方法能通过移动总是成对出现的光标并从显示读数中读取它们的数值从而进行测量。有两类光标:"幅度"和"时间"。使用光标时,要确保将"信源"设置为显示屏上想要测量的波形。要使用光标,可按下 Cursor(光标)按钮。

"幅度"光标 "幅度"光标在显示屏上以水平线出现,可测量垂直参数。"幅度"是参照基准电平而言的。对于数学计算 FFT 功能,这些光标可以测量幅度。

"时间"光标 "时间"光标在显示屏上以垂直线出现,可测量水平参数和垂直参数。"时间"是参照触发点而言。对于数学计算 FFT 功能,这些光标可以测量频率。

"时间"光标还包含在波形和光标的交叉点处的波形幅度的读数。

自动

Measure(测量)菜单最多可采用五种自动测量方法。如果采用自动测量,示波器会为用户进行所有的计算。因为这种测量使用波形的记录点,所以比刻度或光标测量更精确。自动测量使用读数来显示测量结果。示波器采集新数据的同时对这些读数进行周期性更新。

Taking simple measurements

You need to see a signal in a circuit, but you do not know the amplitude or frequency of the signal. You want to quickly display the signal and measure the frequency, period, and peak-to-peak amplitude.

Using Autoset

To quickly display a signal, follow these steps:

1. Push the CH1 MENU button and set the probe option attenuation to 10X.
2. Set the switch to 10X on the P2200 probe.
3. Connect the channel 1 probe to the signal.
4. Push the AUTOSET button.

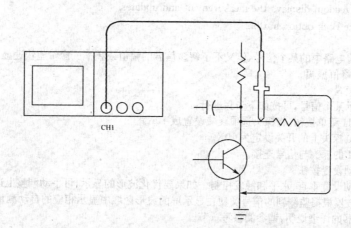

The oscilloscope sets the vertical, horizontal, and trigger controls automatically. If you want to optimize the display of the waveform, you can manually adjust these controls.

NOTE: The oscilloscope displays relevant automatic measurements in the waveform area of the screen based on the signal type detected.

For oscilloscope-specific descriptions, refer to the Reference chapter.

Taking Automatic Measurements

The oscilloscope can take automatic measurements of most displayed signals. To measure signal frequency, period, and peak-to-peak amplitude, rise time, and positive width, follow these steps:

1. Push the MEASURE button to see the Measure Menu.
2. Push the top option button; Measure 1 Menu appears.
3. Push the Type option button and select Freq.

The Value readout displays the measurement and updates.

NOTE: If a question mark (?) displays in the Value readout, turn the VOLTS/DIV knob for the appropriate channel to increase the sensitivity or change the SEC/DIV setting.

4. Push the Back option button.
5. Push the second option button from the top; Measure 2 Menu appears.
6. Push the Type option button and select Period.

The Value readout displays the measurement and updates.

7. Push the Back option button.
8. Push the middle option button; Measure 3 Menu appears.
9. Push the Type option button and select Pk-Pk.

The Value readout displays the measurement and updates.

10. Push the Back option button.
11. Push the second option button from the bottom; Measure 4 Menu appears.
12. Push the Type option button and select Rise Time.

The Value readout displays the measurement and updates.

13. Push the Back option button.
14. Push the bottom option button; Measure 5 Menu appears.
15. Push the Type option button and select Pos Width.

The Value readout displays the measurement and updates.

16. Push the Back option button.

简单测量

您需要查看电路中的某个信号,但又不了解该信号的幅值或频率。您希望快速显示该信号,并测量其频率、周期和峰值振幅。

使用自动设置

要快速显示某个信号,可按如下步骤进行:

1. 按下 CH1 菜单按钮,将探头选项衰减设置成 10X。
2. 将 P2200 探头上的开关设定为 10X。
3. 将通道 1 的探头与信号连接。
4. 按下自动设置按钮。

示波器自动设置垂直、水平和触发控制。如果要优化波形的显示,可手动调整上述控制。

注释:示波器根据监测到的信号类型在显示屏的波形区域中显示相应的自动测量结果。

有关示波器的详细说明,请参阅参考章节。

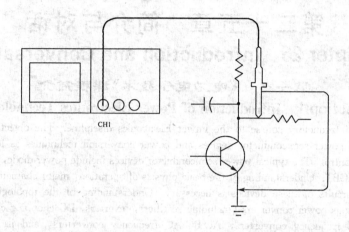

自动测量

示波器可自动测量大多数显示出来的信号。要测量信号的频率、周期、峰峰值、上升时间以及正频宽,可按如下步骤进行:

1. 按下测量按钮,查看"测量菜单"。
2. 按下顶部的选项按钮;显示"测量1菜单"。
3. 按下类型选项按钮,选择频率。

值读数将显示测量结果及更新信息。

注释:如果"值"读数中显示一个问号(?),请尝试将"伏/格"旋钮旋转到适当的通道以增加灵敏度或改变"秒/刻度"设定。

4. 按下返回选项按钮。
5. 按下顶部第二个选项按钮;显示"测量2菜单"。
6. 按下类型选项按钮,选择周期。

值读数将显示测量结果及更新信息。

7. 按下返回选项按钮。
8. 按下中间的选项按钮;显示"测量3菜单"。
9. 按下类型选项按钮,选择峰-峰值。

值读数将显示测量结果及更新信息。

10. 按下返回选项按钮。
11. 按下底部倒数第二个选项按钮;显示"测量4菜单"。
12. 按下类型选项按钮,选择上升时间。

值读数将显示测量结果及更新信息。

13. 按下返回选项按钮。
14. 按下底部的选项按钮;显示"测量5菜单"。
15. 按下类型选项按钮,选择正频宽。

值读数将显示测量结果及更新信息。

16. 按下返回选项按钮。

第二十五章 简介与对话
Chapter 25 Introduction and Conversation

第一节 《电力电子技术》课程简介
Section 1 Introduction of Power Electronics Technology

This is an introductory course to the Power Electronics discipline. The objective is to develop understanding of power semiconductor devices and power conversion techniques as for electric power processing and control. The typical power semiconductor devices include power diode, thyristor, Power MOSFET, and IGBT. Understanding of the basic physics of operation, major characteristics, the drive and protection circuits of these devices is necessary. Understanding of the topologies and operation principles of various power converters (including rectifiers, inverters, DC chopping circuit, AC power control, AC-AC frequency converters, AC-DC-AC frequency converters, and isolated DC to DC switching power supplies) and understanding basic Pulse Width Modulation techniques and soft-switching techniques are the focus parts of this course. Basic circuit calculation and design for rectifier circuits and simple DC-DC converters are also required.

本课程为电力电子学科的入门课程。其目标是培养学生对电力半导体器件和用于电能变换与控制的电力电子变流技术的理解。涉及的典型电力半导体器件包括电力二极管、晶闸管、电力金属氧化物场效应晶体管和门极绝缘双极型晶体管。对这些器件的基本原理、主要特性以及其驱动和保护电路的掌握是必要的。掌握各种电力变换电路(包括整流电路、逆变电路、直流斩波电路、交流电力控制电路、交交变频电路、交直交变频电路和隔离型的直流开关电源电路)的具体电路拓扑和工作原理,以及理解脉冲宽度调制和软开关等控制技术是本课程的核心部分。对相控整流电路和简单直流斩波电路的电路参数计算和设计也是应该掌握的内容。

第二节 矩阵实验室(MATLAB)简介
Section 2 Introduction of MATLAB

MATLAB is a powerful computing system for handling the calculations involved in scientific and engineering problems. The name MATLAB stands for MATrix LABoratory, because the system was designed to make matrix computations particularly easy. MATLAB was originally written to provide easy access to matrix software developed by the LINPACK and EISPACK projects. Today, MATLAB engines incorporate the LAPACK and BLAS libraries, embedding the state of the art in software for matrix computation.

MATLAB has evolved over a period of years with input from many users. In university environments, it is the standard instructional tool for introductory and advanced courses in mathematics, engineering, and science. In industry, MATLAB is the tool of choice for high-productivity research, development, and analysis.

MATLAB is a programming environment for algorithm development, data analysis, visualization, and numerical computation. As a high-performance language for technical computing, it can solve technical computing problems faster than with traditional programming languages, such as C, C++, and Fortran.

Typical uses of MATLAB include:
- math and computation
- algorithm development
- data acquisition
- modeling, simulation and prototyping

第二十五章 简介与对话

- data analysis, exploration and visualization
- scientific and engineering graphics
- application development, including graphical user interface building

MATLAB is an interactive system whose basic data element is an array that does not require dimensioning. This allows you to solve many technical computing problems, especially those with matrix and vector formulations. It would take to write a program in a scalar noninteractive language such as C or Fortran in a fraction of the time.

Toolboxes

MATLAB features a family of add-on application-specific solutions called toolboxes. Very important to most users of MATLAB, toolboxes allow you to learn and apply specialized technology. Toolboxes are comprehensive collections of MATLAB functions (M-files) that extend the MATLAB environment to solve particular classes of problems. Areas in which toolboxes are available include signal processing, control systems, neural networks, fuzzy logic, wavelets, simulation, and so on.

The MATLAB System

The MATLAB system consists of five main parts: development environment, the MATLAB mathematical function library, the MATLAB language, the MATLAB graphics, the MATLAB application program interface (API).

矩阵实验室(MATLAB)是一个解决包括科学和工程计算问题的强大的计算系统。MATLAB的名称源于矩阵实验室的缩写,因为该软件使复杂的矩阵计算变得非常容易。MATLAB软件最初是为了方便进入由 LINPACK 和 EISPACK 项目开发的矩阵软件而开发的。今天,MATLAB 中合并了 LAPACK 和 BLAS 两个库,嵌入到 MATLAB 软件用于矩阵运算。

MATLAB 历经多年的发展演变,吸引了越来越多的遍及各领域使用者。在学校里,它是一个标准的初、高级课程的教学工具,应用于数学、工程和科学研究的各个方面。在工业生产中,MATLAB 也是一个用于高生产率的研究、开发和分析的工具。

MATLAB 拥有用于算法开发、数据分析、可视化和数值计算的程序设计环境。作为一种用于专业技术计算的高级语言,MATLAB 比传统的 C、C++和 Fortran 有着更快的计算速度。

MATLAB 典型的应用包括:
- 数学和计算
- 算法开发
- 数据采集
- 建模、仿真和系统原型的构建
- 数据分析、研究和可视化
- 科学和工程图形的绘制
- 应用程序的开发,包括图形用户口界面的创建

MATLAB 是一个交互式的系统,其基本数据元素是一个不需要标注数组。这可以解决很多专业技术计算问题,尤其是那些矩阵和向量表达式。它可在短时间内,像 C 或 Fortran 一样,用标量的非交互式的语言编写程序。

工具箱

MATLAB 的显著特点是它的工具箱。这些工具箱把各种特定应用领域解决方案组合在一起。这对大多数 MATLAB 用户来说是非常重要的。工具箱可以让你学习和运用某一领域的特殊专业技术。工具箱也是由许多 MATLAB 函数组成的集合。这些 MATLAB 函数(M 文件)扩展了 MATLAB 环境,以解决问题的特定类别。工具箱所涉及的常见领域包括信号处理、控制系统、神经网络、模糊逻辑、小波分析、仿真等等。

MATLAB 系统

MATLAB 系统有 5 个主要组成部分,分别为 MATLAB 的开发环境、MATLAB 的数学函数库、MATLAB 语言、MATLAB 图形、MATLAB 的应用程序界面(API)。

第三节 实践项目简介
Section 3 Introduction of the Practical Project

Good morning, ladies and gentlemen!

May I have a brief introduction of my graduation project: The Design of Soft Starter of Induction Motor.

As we know, squirrel-cage induction motor is applied extensively in driving machines because of its low cost, steady performance and little maintenance. In non-speed regulation case, the start and protection of the AC motor are the main tasks of the electrical control system. The task of motor starting system is to reduce the starting current and time. For the squirrel-cage induction motor, the method of reducing starting current is to decrease the voltage of power supply. For example, connecting reactor or autotransformers, or Y-Δ starting mode is used traditionally. The main protection contents are over current, overload and phase-lacking protection. In case of a breakdown, the protection system cuts off the power supply automatically.

There is a single chip microprocessor in my system acting as the core of system. It is functioning by the thyristor in the main circuit to control the motor's starting and protecting. The structure of the system is shown in Figure 1.

The voltage detecting and current detecting module is designed in the system and they convert the current and voltage to the suitable value and transmit them to the microprocessor. The calculating result of the microprocessor, that is the trigger point of thyristor, is sent to the thyristor through the driver, thus controlling the output voltage. The simple keyboard and digital display unit are designed for setting and displaying parameters.

In simple words, a soft starter limits starting current by means of regulating the voltage of the motor. There are some ways to start motor and the Current Amplitude Limiting is chosen in my system. In this mode, the starting current is limited to the setting value I_{max}. In fact, the system is functioning as a closed-loop system of the follow-up control with the regulated variables of the thyristor voltage and motor current. The PID regulator is used in the system and the starting current curve is regulated according to the given curve as shown in Figure 2.

In addition to the controlling function of the starting of the motor, the system can also prevent the motor from over-current, overload and lacking phase. That is, when the current of motor exceeds the set value, when the motor runs with the current over normal current I_N for a certain period of time, or when the 3-phase is not full, the system cuts off the power to the motor and output the alarm simultaneously. The system indicates the parameters such as current, voltage and power factors when the motor runs normally.

I have tested the system on a squirrel-cage induction motor of 2 kilowatt and the satisfactory result has been achieved.

That is all.

Thanks!

女士们,先生们,大家早上好!

请允许我简单地介绍一下我的毕业设计题目:"感应电动机软启动器设计"。

众所周知,鼠笼式感应电动机由于其价格低廉、性能稳定、维护量小等优点被广泛应用于机械设备的拖动。在无调速的场合,交流电动机的启动和保护是电气系统设计要考虑的主要问题。电动机启动系统的任务是降低启动电流,缩短启动时间,而对于鼠笼式感应电动机来讲,降低启动电流的手段即是降低电动机的供电电压。传统的起动方式有串电抗器起动、自耦变压器起动,星三角起动等方式;而电动机的保护内容主要有过流保护、过载保护、缺相保护等。上述故障发生时,系统自动切断电源。

第二十五章 简介与对话

我所设计的交流电动机软启动器是以单片机为核心,通过对设置在主回路上的晶闸管的控制,实现了交流电动机启动和保护的基本功能。系统组成如图1所示。

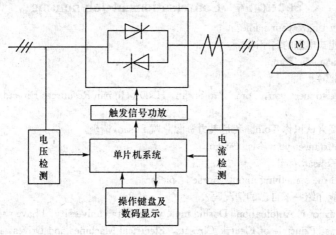

图1 电动机软启动器结构图

系统中设有电压电流检测模块,负责将电压电流转换成合适的量值送入单片机,单片机的计算结果,即晶闸管的触发角,经功率放大后送给晶闸管,控制其输出电压。为了进行参数设置和显示,系统中设有简单的键盘和数码管显示装置。

简单地讲,电动机软启动器是以调节电机电压为手段来限制启动电流的装置。目前,启动方式有很多种,我所选择的启动方式是电流限幅起动方式,此方式使启动电流电流限制在所设定的最大电流以内。事实上,本系统是一个以电流为被调量,以晶闸管电压输出为调节量的闭环随动控制系统,调节器为 PID 调节器,系统的调节目标即是使电动机的启动电流按照事先设定好的曲线运行,如图2所示。

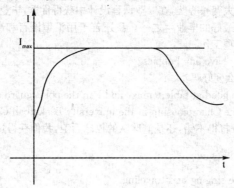

图2 启动电流曲线

本系统除具有电动机启动控制功能以外,还具有电动机的过流保护、过载保护和缺相保护功能,即当电动机电流超过设定的最大限值时或者超过额定限值并持续运行一定时间时,或者三相电缺相时,系统将停止向电动机供电,同时输出报警。电动机正常工作时,系统将显示工作电流、电压、功率因数等参数。

本系统针对一个 2kW 的鼠笼式感应电动机进行了测试,得到了基本满意的测试结果。

就介绍到这里,谢谢!

第四节 就业情景对话
Section 4　Conversations at Job-hunting

——Susan：May I have your name?

——Susan：我能知道您的名字吗?

——Tom：My name is Tom.

——Tom：我叫汤姆。

——Susan：Nice to meet you, Tom. I'm Susan, HRD (Human Resources Director) of the company. Please have a seat.

——Susan：很高兴认识你 Tom。我是人力资源总监 Susan，请坐。

——Tom：Nice to meet you, Ms. Susan.

——Tom：您好 Susan 女士。

——Susan：Tell me something about yourself, please.

——Susan：您能介绍一下自己的情况吗?

——Tom：My major is Automation. During my 4 years in the university, I have passed all the courses with honors such as: Principle of Electric Circuits, Electrical Machines and Drives, Power Electronics, Automatic Control Theory, Process Control System, and so on. Although I'm not the best one in my class with regard to the grades, I have my advantages. I won the first prize at the Design Contest of Micro-computer Products. I also got a grade A in major course design. I got Excellent in graduation thesis. So you can see I am absolutely one of the best students in terms of experimental and practical abilities. I am also good at programming both with assembly language and advanced languages such as C++ and VB. So I think I'm suitable for this job.

——Tom：我的专业是自动化。在大学的四年中，我以优异成绩通过所有专业课，如：电路原理、电机与拖动、电力电子、自控原理、过程控制系统等。尽管按各科成绩我在班级不是最好的，但我有自己的优势。我曾赢得微机产品设计大赛一等奖，在专业课程设计中获得优秀，毕业设计论文也是优秀。你可以看出在动手能力上我绝对是最好的毕业生之一。我还善于用汇编语言和高级语言如 C++ 和 VB 编程。因此我觉得我很适合这项工作。

——Susan：Good. Do you have any hobbies?

——Susan：好的，你有什么爱好吗?

——Tom：Of course. I'm good at table-tennis, and I'm the point guard of the Basketball Team in the department. We once won the Championship in the university Basketball Contest.

——Tom：当然，我乒乓球打得不错，还是篮球队的控球后卫，我们获得过学院冠军。

——Susan：Really?

——Susan：真的吗?

——Tom：Yes, I also like singing and traveling.

——Tom：是的，我还喜欢唱歌和旅行。

——Susan：That's impressive. From your words, I can see that you performed excellently in your university. What I want to know is why you want to choose this job.

——Susan：很不错。从你的谈话中我可以看出你在校时很优秀。我想知道为什么你想到这里来工作。

——Tom：I have long been keen on the news from your industry, and deeply impressed by the development of your company. As I know, TMD Company is a joint venture and listed the top 5 in this field in China. In the past several years, many excellent graduates were recruited to your company. I have talked with some of them, so I know something about your company. I also admire and agree with the employee development values that the company has adopted. I find that my personal goals and ideas about business operations fit perfectly with the company's. I am confident that I will have a very rewarding and successful career here.

——Tom：我一直关注你们行业的新闻，贵公司的发展给我留下了很深的印象。据我所知TMD公司是一个合资企业，在国内位于该领域前五。在过去几年中，有很多优秀的大学毕业生来到你们公司。我和他们中的一些人交谈过，对你们公司有一些了解。同时我也很喜欢贵公司培养员工的企业文化。我觉得我的个人发展目标、商业理念和公司的目标完全吻合。我觉得我在这儿能开创有价值而且成功的事业。

——Susan: Commitment and teamwork are very important to a company. Do you have any ideas about that?

——Susan：员工的贡献精神和团队精神对一个公司是非常重要的。关于这一点，你有什么想法？

——Tom: I am very responsible. My teacher and classmates would always count on me to complete a certain task. There is also no need to remind me of what I should do. I enjoy doing things and cooperating with others.

Tom：我很有责任心，如果让我负责一件事，老师和同学们都会很放心。而且我用不着别人提醒我该做什么。我喜欢与人合作一起做事情。

——Susan: So what are your career objectives? And how do you perceive your own development if you were to assume this position?

——Susan：那你的事业目标是什么？如果你获得了这个职位，你对自己的发展有什么想法？

——Tom: I expect to have a good opportunity to put all of my knowledge into practice. I'd like to start with the research and development work. I am a doer and believe I can contribute a great deal to the company. I hope that within a couple of years, I could manage my work independently and were able to lead an energetic and productive team.

——Tom：我希望得到很好的实战机会。我想从研发工作做起，我是个实干者，我相信我能为公司贡献很多。我希望在几年内，我能够独立承担自己的工作并且领导一支有活力及高效率的团队。

——Susan: Good, then do you have any questions about our company?

——Susan：好的，对于公司，你有什么问题吗？

——Tom: Could you please tell me something about your training program?

——Tom：能否介绍一下你们的培训制度？

——Susan: In brief, we offer both in-house and off-site training to US headquarters. We have a few daylong training sessions for topics like business writing skills and software training, etc. These sessions are available to everyone who applies. We also have a variety of other programs based on each work function. Our program is essentially a job-rotation program, and we believe it is more effective than traditional on-the-job training.

——Susan：简单来说，我们既有在岗培训，也有去美国总部专门培训。我们采用整日课程来进行商务写作和软件操作一类的培训。每个人都可以申请参加。我们还针对各个职能安排了多种多样的课程。我们的课程主要是轮岗，我们相信这比传统的在职培训要有效得多。

——Tom: It sounds attractive!

——Tom：听上去真的很吸引人啊！

——Susan: Lastly, what's your main consideration in your current job-hunting?

——Susan：最后，你选择工作时主要考虑哪个方面？

——Tom: Potential growth opportunity. If a potential job does offer me an opportunity to grow and become mature, I will find it very easy to devote myself to executing my responsibilities at my best. I have reasons to believe that your company will be a perfect place for me to grow and develop!

——Tom：潜在的发展机会。如果一个工作能够让我有发展和成熟的机会，我就会更容易地全身心投入去完成我的职责。我有理由相信贵公司将是我成长和发展的良好平台！

——Susan：I also hope to have the chance to get to know you better in the future.

——Susan：好的，我也希望有机会更加详细地了解你。

——Tom：Thank you for taking your time to talk whith me. I am looking forward to having an opportunity to work here.

——Tom：谢谢您花费时间与我交谈。我期待着有机会来这里工作。

——Susan：You'll hear from us by next week.

——Susan：我们下周会通知你。

——Tom：Thank you very much. See you then.

——Tom：非常感谢您！再见。

参考文献

[1] 徐科军,马修水,李晓林,等. 传感器与检测技术. 北京:电子工业出版社,2011.
[2] 陈建元. 传感器技术. 北京:机械工业出版社,2008.
[3] 唐文彦. 传感器. 北京:机械工业出版社,2007.
[4] 胡向东,刘京诚,等. 传感技术. 重庆:重庆大学出版社,2006.
[5] 王俊杰. 检测技术与仪表. 武汉:武汉理工大学出版社,2009.
[6] Ajay V Deshmukh . Microcontrollers:Theory and Applications. McGraw-Hill. 2005,1. James W. Stewart. The 8051 microcontroller:hardware, software, and interfacing. Prentice Hall,1998.
[7] 江太辉,石秀芳. MCS-51系列单片机原理与应用. 广州:华南理工大学出版社,2002.
[8] 李发海. 电机与拖动基础. 3版. 北京:清华大学出版社,2005.
[9] Theodore Wildi. Electrical Machines. Drives and Power Systems. Fifth Edition. 北京:科学出版社, 2002.
[10] A. E. Fitzgerald. Electric Machinery. Sixth Edition. 北京:清华大学出版社,2003.
[11] 王兆安. 电力电子技术. 北京:机械工业出版社,2007.
[12] Timothy J. Maloney. Modern Industrial Electronics. 北京:科学出版社,2002.
[13] 邱关源. 电路. 北京:高等教育出版社,2006.
[14] Charles K. Alexander and Matthew N. O. Sadiku Fundamentals of electric Circuits. First edition, Tsinghua University Press.
[15] James W. Nilsson. Susan A. Riedel. Electric Circuits (Sixth Edition). 北京:科学出版社,2003.
[16] L. A. Bryan and E. A. Bryan PROGRAMMAL CONTROLLERS THEORY AND IMPLEMENTATION. Second Edition, 1997. By Industrial Text Company Published by Industrial Text Company Printed and bound in the United States of America.
[17] AutoCAD Electrical 2010 Getting Started. Part No. 225B1-050000-PM01A. Published by Autodesk, Inc. 2009.
[18] 阳宪惠. 工业数据通信与控制网络. 北京:清华大学出版社,2003.
[19] John Park et al. Practical Data Communications for Instrumentation & Control. Elsevier. 2003.
[20] 刘秀玲,黄建兵. 集散控制系统. 北京:中国林业出版社,2006.
[21] http://en. wikipedia. org/wiki/Distributed_control_system
[22] 潘新民,王艳芳. 微型计算机控制技术. 北京:电子工业出版社,2011.
[23] 高国琴,等. 微型计算机控制技术. 北京:机械工业出版社,2006.
[24] 于海生,等. 微型计算机控制技术. 北京:清华大学出版社,2009.
[25] 冯培悌. 计算机控制技术. 杭州:浙江大学出版社,2005.
[26] 高金源,夏洁. 计算机控制系统. 北京:清华大学出版社,2007.
[27] Astrom K. J. and Writtenmark B. . Computer Controlled Systems Theory and Design. 北京:清华大学出版社,2002.
[28] 裘祖荣. 精密机械设计基础. 北京:机械工业出版社,2011.
[29] 刘德全. 语码转换式双语教学系列教材机电一体化技术. 大连:大连理工大学出版社,2008.
[30] 徐国凯. 语码转换式双语教学系列教材电子与自动化技术. 大连:大连理工大学出版社,2008.
[30] 陈虹. 楼宇自动化技术与应用. 北京:机械工业出版社,2005.
[32] Drew Gislason. ZigBee Wireless Networking. Elsevier Inc.
[33] Ian F. Akyildiz. Wireless Sensor Networks. A John Wiley and Sons, Ltd, Publication,2010.
[34] 阮毅. 电力拖动自动控制系统——运动控制系统. 4版. 北京:机械工业出版社,2009.

[35] Bimal K. Bose. Modern Power Electronics and AC Drives. 北京:机械工业出版社,2003.
[36] 费业泰. 误差理论与数据处理. 北京:机械工业出版社,2010.
[37] 程德福,林君. 智能仪器. 北京:机械工业出版社,2005.
[38] Richard C. Dorf, Robert H. Bishop, Modern Control Systems. Tenth ed. English reprint Copyright C 2005 by Science Press and Pearson Education Asia Limited.
[39] Francis H. Raven. Automatic Control Engineering. Third edition. McGraw-Hill Book Company, 1978.
[40] Palm William John. Control systems engineering copyright by John Wiley & Sons, Inc. 1986.
[41] http://www.mathworks.cn
[42] http://pt.csust.edu.cn/eol/homepage/course
[43] Brian Hahn, Daniel T. Valentine. ESSENTIAL MATLAB For Engineers and Scientists. Third Edition. Typeset by Charon Tec Ltd (A Macmillan Company), Chennai, India, Printed and bound in Italy,2007.
[44] TDS1000 和 TDS2000 系列数字存储示波器用户手册. 泰克科技(中国)有限公司,2005.
[45] Charles K. Alexander and Matthew N. O. Sadiku, Fundamentals of electric circuits. Tsinghua University Press.
[46] Thomas L. Floyd. Digital Fundamentals. 北京:科学出版社,2002.
[47] Thomas L. Floyd. 数字基础. 李晔,等,译. 北京:清华大学出版社,2005.
[48] 康华光,陈大钦,张林. 电子技术基础模拟部分. 5 版. 北京:高等教育出版社,2009.
[49] Allan R. Hambley. 电子技术基础(英文改编版). 北京:电子工业出版社,2007.

索 引

(电流)断续模式 4-4
(电流)连续模式 4-4
0/1 序列 9-1
0 型系统 20-5
16 位机 2-1
1 型系统 20-5
2 型系统 20-5
3σ 准则 16-2
4 极汽轮发电机 3-9
4 位机 2-1
4 线制传感器 6-3
555 定时器 21-7
8 位机 2-1
AA 电池 15-2
ABC/或正/相序 5-12
ACB/或负/相序 5-12
ActiveX 控件脚本程序 8-1
AD6 网络升级 17-1
Altium 公司印刷电路板设计软件,版本号 6.x,简记:AD6 17-1
ASCII 码 2-1
ASCII 协议 9-6
A 类评定 16-4
A 相电压 5-12
BCD 码 2-1
BTL 电路 22-8
B 类评定 16-4
B 相电压 5-12
CAM 编辑器 17-1
CCD 图像传感器 1-5
CMOS 电路 22-5
C 相电压 5-12
DB 编辑器 6-4
DIN 导轨 DIN 6-3
DP 标识符 6-6
DP 标准从站 6-6
DP 从站接口 6-6
DP 延迟时间 6-6
DP 主站系统 6-2

D 锁存器 21-5
EPROM 编程器 2-1
F 分布 16-2
h 参数模型 22-4
h 参数小信号模型 22-4
I/O 总线网络 6-6
IC 卡 14-5
LCD 显示器 19-3
L 端子 6-3
MAC 地址 15-3
Modbus 协议 9-6
MOS 控制晶闸管 4-1
MPI 卡 6-6
NFF 型温度传感器 1-3
N 极 3-6
N 型 22-3
PD 控制 20-7
PID 控制 20-7
PID 控制算法 10-2
PI 调节器 18-1
PI 控制 20-7
PROFIBUSDP 主站 6-6
PROFIBUS 总线连接器 6-6
P 型 22-3
RC(电阻-电容)电路 5-7
RL(电阻-电感)电路 5-7
RLC 并联电路 5-7
RLC 串联电路 5-7
RLO 沿检测 6-4
SCL 编辑器 6-4
SPWM 逆变器 18-6
S 极 3-6
s 平面 12-7
s 域的等效电路 5-13
TTL 兼容 2-2
t 检验法 16-2
T 形网络 5-10
T 型电阻解码网络 2-8
T 型反馈网络 22-2
U_{BE} 倍增 22-4

u-i 特性 5-1
V 带传动 13-7
X 方向间隔 17-7
Y-Y 电路 5-12
Y 方向间隔 17-7
ZigBee 规范 15-2
ZigBee 联盟 15-2
ZigBee 设备对象 15-4
ZigBee 设备描述 15-4
z 变换 12-7
z 平面 12-7
Δ—Y 变换 5-2
χ^2 分布 16-2

A

阿基米德蜗杆 13-6
安防自动化 14-1
安排(装置) 13-1
安培 5-1
安全 8-1
安全工作区 4-1
安全间距 17-5
安全离合器 13-8
安全区保护 22-10
安全系数 4-5
安全销 13-8
安全装置 10-4
安全子域 15-5
安匝数 3-5
安装 8-1
鞍点 20-10
按键 2-7
按键脚本程序 8-1
按钮 6-1
按照 3-5

B

八进制 2-1
巴特沃思滤波器 22-2
白炽灯 3-3
摆动从动件 13-7
摆线齿轮 13-5

摆线运动规律 13-5
板层摆放管理 17-5
版本 8-1
版本控制 17-1
版权 8-1
办公自动化系统 14-1
半波整流 22-10
半导体 22-3
半导体二极管 22-3
半导体基本概念 22-3
半导体激光器 1-5
半导体气体传感器 1-4
半导体应变式压力传感 1-2
半导体整流器 4-1
半加器 21-4
半径 17-2
半控型器件 4-1
半桥电路 4-2
半桥整流电路 18-1
半双工 2-6
半双工方式 12-5
半圆键 13-8
半圆铭牌 7-4
帮助系统 7-1
绑定 15-4
包交换 9-7
包角 13-7
包络 13-6
饱和 3-1
饱和电流 21-3
饱和电压 18-2
饱和非线性控制 18-2
饱和区 22-4
保持 19-2
保持、有效 21-6
保持架 13-9
保持时间 21-5
保存设计空间 17-2
保护电路 4-1
保护现场 2-4

保护装置 10-4
保留符号 2-3
保险丝熔断 6-1
报警 8-1
报警查询 8-1
报警程序设计 12-4
报警处理 4-1
报警传感器 14-5
报警电路 4-1
报警确认 8-1
报警系统 14-5
报警限 8-1
报警装置 14-5
贝克箱位电路 4-2
贝塞尔公式 16-2
贝塞尔曲线 17-2
备份 17-1
备用电池 6-3
备用电源 14-4
备用端子 7-4
备择路径 6-5
背板总线 6-3
背对背二极管 4-1
背对背配置(滚动轴承) 13-9
背光 2-7
背锥 13-6
倍频 2-5
倍频器 22-9
倍压整流电路 22-8
倍增器 19-1
被测变量 10-3
被测参数 12-1
被动式红外线传感 14-5
被加数 21-4
被减数 21-4
被控参数 12-7
被控对象 20-1
被控制量 20-1
本征 22-3
泵 3-4
比较电压 18-1
比较机构 20-1
比较型模数转换器 19-2
比例电流源 22-6

比例度 10-3
比例环节 20-2
比例积分控制器 6-7
比例积分微分(PID)控制 18-1
比例控制 18-1
比例控制器 6-7
比例因子 17-2
比例运算电路 22-2
比例增益 10-2
比特率 21-1
比值控制 10-4
必要条件 20-3
闭合回路 21-6
闭合曲线 20-5
闭环 18-1
闭环传递函数 20-2
闭环放大倍数 18-1
闭环极点 18-2
闭环控制系统 20-1
闭环零点 18-2
闭环脉冲宽度调速系统 12-4
闭环增益 22-7
闭路电视监视系统 14-5
闭式链 13-3
边界空白区域的宽度 17-5
边界线宽度 17-5
边沿 2-5
边沿触发 2-4
边缘效应 1-2
编程语言 6-2
编号索引控制 17-3
编辑规则属性 17-6
编辑元件 7-3
编码 19-5
编码键盘 19-3
编码器 2-2
编译 15-6
编译集成元件库 17-4
编址 2-2
变比 3-5
变风量 14-3
变化率报警 8-1
变换特性 1-2

变跨导型乘法器 22-4
变量 2-3
变量表 2-3
变量声明表 6-4
变量值保存 8-1
变流技术 4-2
变频电动机 18-5
变频调速 18-5
变频器 3-8
变送器 10-2
变速控制 12-4
变位齿轮 13-6
变系数 10-3
变压变频 4-8
变压器 5-10
变压器反馈式 22-9
变压器反馈式正弦波振荡电路 22-9
变压器漏抗 18-1
变压器耦合 22-4
变压器耦合放大电路 22-4
标称传递函数 20-7
标称电阻 1-3
标称值 3-4
标定 1-5
标定量程 1-5
标度变换 19-4
标度变换原始值 8-1
标度转换 12-6
标识符字典 8-1
标题框和刻度栏 17-5
标幺值 3-5
标注线宽度 17-5
标准不确定度 16-4
标准差 16-2
标准差的合成 16-3
标准齿顶高 13-6
标准齿高 13-6
标准齿轮 13-6
标准单元 19-1
标准电信号 12-1
标准公差 13-10
标准公差等级 13-10
标准和协议 9-1
标准化 13-10

标准压力角 13-6
标准总线 12-1
表面传输阻抗 14-6
表面粗糙度测量仪器 13-10
表面粗糙度的测量 13-10
表面热处理 13-2
表面声波式触摸屏 19-3
表示层 14-2
表述层 9-5
别捷尔斯法 16-2
丙类 22-8
并励 3-3
并联 3-5
并联电路 5-3
并联电路分析 5-2
并联连接的电路元件 5-2
并联谐振 5-11
并联谐振式逆变电路 4-7
并联谐振阻抗 5-11
并联型 22-10
并入串出 21-6
并行程序设计法 12-8
并行加法器 21-4
并行数据 2-6
并行通讯 2-6
并行总线 12-5
波动 10-4
波绕组 3-3
波特 2-6
波特率 2-6
波特率发生器 2-6
波特图 22-4
波纹 12-8
波纹管 1-7
波纹管联轴器 13-8
波形 2-2
波形编辑器 17-1
波形变换电路 22-9
波形发生电路 22-9
波形整形 22-1
玻璃液面计 10-4
伯德图 20-5
铂电阻 1-3
薄膜开关 17-5
补偿控制 18-1

索 引

补偿器 20-6
补偿绕组 3-4
补码 2-1
捕捉栅格 17-2
不变的系统误差 16-2
不对称三相电路 5-12
不挥发存储器 2-1
不间断电源 4-8
不具有电气意义的线 17-1
不可控器件 4-1
不可逆 PWM 变换器 4-3
不可重复触发 21-7
不控整流 4-2
不平衡电容 1-2
不确定度的报告 16-4
不同公式计算标准差比较法 16-2
布尔代数 21-2
布尔运算 2-3
布局 17-5
布线 17-5
布线方案 17-5
布线拓扑类型 17-6
布线完成率 17-5
布线优先次序 17-6
步进电动机 3-10
步骤启用 6-5
部分分式 20-8
部分分式法/分解定理 5-13

C

擦除 21-9
采集 19-2
采样 2-8
采样保持电路 12-2
采样保持器 2-8
采样抽取技术 19-2
采样定理 12-2
采样开关 12-8
采样控制系统 20-8
采样率 21-10
采样频率 12-2
采样数据系统 12-8
采样误差 19-2
采样系统 12-2

采样周期 12-2
菜单 8-1
参考电平 12-2
参考电压 2-8
参考方向 5-1
参考结点 5-3
参考输入 20-1
参考相量 5-8
参量 20-4
参数 17-2
参数配合 18-2
参数优化 20-10
参数整定 12-7
残液 10-4
残余误差观察法 16-2
残余误差校核法 16-2
操纵 22-1
操作成本 10-4
操作过电压 4-1
操作码 2-3
操作数 2-3
操作系统 8-1
操作员系统标识符 8-1
操作员站 11-2
操作站 12-1
操作指导 12-1
槽 3-3
槽销 13-8
测控系统 9-1
测量 10-2
测量不确定度 16-4
测量不确定度的合成 16-4
测量单位 8-1
测量电路 1-1
测量方程 16-5
测量技术 19-1
测量信号 10-4
测量元件 20-1
测量值 12-7
测试点 17-6
测速发电机 18-1
测温电路 1-3
层次设计 17-2
插入/编辑辅元件 7-3
插入/编辑元件 7-3

插入导线 7-2
插入附件 7-4
插入示意图 7-4
插入线号 7-2
查表技术 12-6
查询 2-5
查询方式 2-5
查找相类似对象 17-3
差 21-4
差错检测 9-2
差错控制技术 12-5
差动变压器 1-2
差动电容传感器 1-2
差动轮系 13-6
差动输入 5-5
差动输入电压 5-5
差分对布线 17-6
差分方程 20-8
差分放大电路 22-6
差分输入 22-2
差复励发电机 3-3
差模放大倍数 22-6
差模信号 22-6
差压 1-7
差压变送器 1-2
差压传感器 1-2
差压式流量计 1-7
差压指示器 1-2
掺杂 22-3
常闭 6-2
常闭触点 6-1
常规控制点 11-3
常开触点 3-8
常数 2-3
常数表 2-3
常系数 10-3
常用配合 13-10
场控晶闸管 4-1
场效应管 4-1
超驰控制 10-4
超大规模集成电路 21-4
超导体 5-1
超调 12-8
超调量 20-3
超前 5-8
超前角 4-2

超前进位产生器 21-4
超前进位加发器 21-4
超前校正 22-4
超前滞后环节 10-4
超越离合器 13-8
撤销/重做 17-2
衬底 22-5
成对安装(滚动轴承) 13-9
成分检测 1-7
城域网 9-7
乘法器 2-1
乘法摄动 20-7
乘法运算电路 22-2
乘积项 21-2
程控增益放大器 19-2
程序/流程控制指令 6-4
程序存储器扩展 2-2
程序和顺序控制 12-7
程序结束 2-3
程序流程图 12-10
程序名 2-3
程序模板 11-3
程序判断滤波 12-6
程序区 2-2
程序执行 6-2
迟滞 1-1
尺寸 13-10
尺寸公差(公差) 13-10
尺寸系列(滚动轴承) 13-9
齿顶 13-6
齿顶厚 13-6
齿顶压力角 13-6
齿顶圆直径 13-6
齿顶圆半径 13-6
齿根高系数 13-6
齿根厚 13-6
齿厚 13-6
齿廓工作段 13-6
齿廓啮合基本定律 13-6
齿轮 3-2
齿轮范成原理 13-6
齿轮系 13-6
齿轮形插刀 13-6
齿面接触 13-6

齿式联轴器 13-8
齿条形插刀 13-6
充电 21-7
充分条件 20-3
充要条件 20-3
冲击电压保护 4-3
冲激函数 5-7
冲激响应 5-7
重布 17-5
重复性 1-1
重根 20-3
重击 12-3
重极点 5-13
重新打开上一次的工作空间 17-1
重整装置 10-4
抽头 3-8
出错信号 2-7
出入口控制系统 14-5
出射角 20-4
初级电路/原边 5-10
初级线圈 5-10
初级智能 19-1
初拉力 13-7
初始电流 5-13
初始电平 22-1
初始电压 5-7
初始化 2-2
初始条件 5-13
初始值 2-5
初始转速 3-4
初始状态 10-3
初相 5-8
初值 5-7
初值定理 5-13
除法器 2-1
除法运算电路 22-2
储能 4-3
储能元件 5-6
处理阶段 6-2
触点 6-4
触发 2-4
触发角 4-2
触发器 2-2
触发延迟角 4-2
触发装置 18-1

触摸屏 19-3
触摸屏技术 19-3
穿越频率 20-5
传导 3-8
传递函数 20-2
传动比 13-7
传动角 13-4
传动特性 13-4
传动装置 13-7
传感器 10-2
传感器信号 6-1
传输层 9-5
传输控制协议/网际协议 6-6
传输门 21-3
传输特性 21-3
传输线阻抗 5-12
传输延迟时间 21-3
传统仪器 19-1
串并联(文氏桥)式振荡器 22-9
串并联电路 5-2
串并联电路分析 5-2
串级控制 10-4
串励 3-3
串联 3-5
串联电路 5-3
串联电路分析 5-2
串联连接的电路元件 5-2
串联校正 20-6
串联谐振 5-11
串联谐振阻抗 5-11
串联型 22-10
串联型稳压电路 22-10
串行 2-6
串行程序设计法 12-8
串行端口通信 6-6
串行接口 2-6
串行进位加法器 21-4
串行口中断 2-4
串行输入 21-5
串行数据 2-6
串行数据通信 9-3
串行通信 12-5
串行通信标准总线 12-5
串行通讯 2-6

串行总线 19-3
窗口脚本程序 8-1
创建新图形 7-1
垂直尺寸 8-1
垂直等间距排列 17-3
垂直度 13-10
垂直母线 7-2
纯滞后时间 10-3
磁场 5-10
磁场强度 3-1
磁畴 3-1
磁导率 3-1
磁电特性 1-4
磁化曲线 3-1
磁极 3-3
磁卡 14-5
磁力线 3-3
磁链 5-10
磁路 3-1
磁密 3-1
磁敏传感器 1-4
磁敏二极管 1-4
磁敏三极管 1-4
磁耦合 5-10
磁盘驱动器 2-2
磁通 3-1
磁通密度 3-1
磁通扭曲 3-4
磁通势 3-1
磁性材料 3-1
磁栅传感器 1-6
磁致伸缩式传感器 1-4
磁滞 10-2
磁滞回线 3-1
磁滞损耗 3-1
磁阻传感器 1-4
次级电路/副边 5-10
次级线圈 3-5
次态 21-6
次要带 20-8
从动带轮 13-7
从动件 13-5
从动件滚子 13-5
从组合中解除对象 17-3
聪敏 19-1
粗大误差 1-1

簇 15-3
催化剂 10-4
催化裂化装置 10-4
淬火 13-2
萃取塔 10-4
存储 2-1
存储单元 21-8
存储电路 21-6
存储矩阵 21-8
存储器 2-1
存储器读信号 2-2
存储器扩展 2-2
存储器容量 2-2
存储器写信号 2-2
存储时间 21-8
存储状态字的溢出位 6-4
存在性定理 20-9
存滞后 12-8
错误信息 2-3
错误组织块 6-4

D

达林顿复合管 22-8
打印机 2-1
打印头 2-1
大功率场效应管 12-4
大规模集成电路 21-4
大林算法 12-8
大型计算机 2-1
代数插值 19-4
代数和 5-1
代数形式 5-9
带长 13-7
带传动 13-7
带宽 5-11
带通滤波器 22-9
带阻滤波器 22-9
戴维宁等效电路 5-4
戴维宁定理 5-4
单变量系统 12-9
单刀开关 3-8
单点传送 15-3
单调性 21-10
单端电路 4-8
单工、半双工和双工 9-2
单工方式 12-5
单回路反馈控制 10-3

索引

单级行星轮系 13-6
单极点 5-13
单极型集成电路 22-6
单极型晶体管 22-5
单列轴承 13-9
单面印制板 17-5
单片机 2-1
单片机系统 12-10
单片集成稳压器 22-10
单输入单输出 12-9
单头蜗杆 13-6
单位冲激函数 5-7
单位加速度信号 12-8
单位阶跃 10-2
单位阶跃函数 5-7
单位阶跃响应 20-3
单位阶跃信号 12-8
单位速度信号 12-8
单稳态触发 21-7
单相 3-5
单相半波调速系统 4-2
单相半波可控整流电路 4-2
单相半桥逆变电路 4-2
单相等效电路 5-12
单相桥式全控整流电路 4-2
单相全波可控整流电路 4-2
单相全桥逆变电路 4-2
单向可控硅 12-4
单向离合器 13-8
单向推力轴承 13-9
单一要素 13-10
当量齿轮 13-6
当量载荷 13-9
当前报警 8-1
当前图层 17-5
当前文档中的所有对象 17-3
导出 8-1
导出规则 17-6
导杆 13-4
导航 17-1
导抗 5-14
导轮 13-7

导纳 5-9
导纳参数 5-14
导纳参数矩阵 5-14
导纳三角形 5-9
导入 8-1
导入规则 17-6
导体 3-1
导体数 3-4
导条 3-7
导通角 4-2
导线编号 7-2
导线布线 17-2
导线段 7-2
导线面 17-5
导线图层 7-2
导向平键 13-8
倒向比例电路 5-5
倒向放大器 5-5
倒向输入端 5-5
倒置梯形 21-10
登陆 8-1
等百分比阀 10-2
等臂电桥 1-2
等待请求 12-5
等加速一等减速运动规律 13-5
等价状态 21-6
等速运动规律 13-5
等效变换 5-9
等效串联电阻 5-2
等效导纳 5-9
等效电导 5-2
等效电抗 5-9
等效电路 5-4
等效电纳 5-9
等效电阻 5-2
等效发电机 5-4
等效逻辑电路 2-5
等效阻抗 5-9
等腰双曲柄机构 13-4
等值电路 3-3
等值电路 3-5
低电平触发 2-2
低电压纹波 4-5
低复杂度 15-1
低副 13-3

低功耗 15-1
低阶系统 20-3
低频 20-5
低频功率放大电路 22-8
低数据速率 15-1
低通 19-2
低通滤波器 22-9
低通数字滤波 12-6
低温试验 19-6
低选器 10-4
低压边 3-5
低压绕组 3-5
低延迟 15-1
狄克松准则 16-2
底层 17-5
底层丝印层 17-5
底座连接器 6-3
地址标识符 6-3
地址分配列表 6-3
地址缓冲器 2-1
地址锁存 2-2
地址锁存器 2-2
地址锁存信号 2-2
地址锁存允许 12-5
地址信号 2-2
地址帧 2-6
地址总线 2-2
递增-递减计数器 21-6
第三象限 3-8
第一象限 3-8
典型环节 20-2
典型系统 18-2
点对点协议 14-3
点接触型 22-3
点阵式显示器 12-3
电/气转换器 10-4
电磁阀 10-4
电磁法 1-6
电磁干扰 19-6
电磁感应 3-1
电磁功率 3-7
电磁离合器 13-8
电磁流量计 1-7
电磁时间常数 18-1
电磁效应 1-4
电磁转矩 3-7

电导 5-1
电动测微仪 1-2
电动机 3-2
电动机惯例 3-9
电动势 3-6
电动势调节器 18-2
电动执行器 12-1
电度表 3-5
电镀 13-2
电感 5-6
电感的串并联 5-10
电感反馈式(电感三点式) 22-9
电感反馈式正弦波振荡电路 22-9
电感线圈 1-2
电感元件 5-6
电感值 3-5
电功率 3-4
电光效应 1-4
电荷 5-1
电机 3-1
电极栅 1-2
电角度 3-6
电解电容 3-8
电抗 5-9
电可擦除的 PROM 21-8
电缆电视 14-5
电力半导体器件 18-1
电力变换 4-8
电力变流器 18-1
电力场效应晶体管 4-1
电力电子技术 4-8
电力电子器件 4-1
电力电子系统 4-8
电力电子学 4-8
电力二极管 4-1
电力公害 18-1
电力晶体管 4-1
电力拖动 18-1
电力系统 14-3
电力系统分析 4-8
电流 5-1
电流(源)型逆变电路 4-5
电流表 5-1
电流并联 22-7

电流串联 22-7
电流调节器 18-2
电流反馈 18-1
电流反馈系数 18-1
电流方程 22-3
电流方向 5-9
电流放大模型 22-1
电流放大系数 22-4
电流负反馈 18-1
电流互感器 3-5
电流环 18-2
电流基值 3-5
电流截止负反馈 18-1
电流可逆斩波电路 4-3
电流控制电流源 5-4
电流控制电压源 5-4
电流连续模式 4-3
电流输出端 2-8
电流相量 5-8
电流源 5-4
电流源逆变器 18-6
电流增益 22-1
电流正反馈 18-1
电路 5-1
电路板布线规则 17-5
电路板查看显示 17-5
电路板选项 17-7
电路参数 5-1
电路仿真 17-2
电路分析 5-2
电路符号 5-1
电路交换 9-7
电路理论 5-2
电路模型 5-1
电路图 5-1
电路选择性 5-11
电路元件 5-1
电路元件方程的相量形式 5-8
电纳 5-9
电平 2-5
电平触发 2-4
电平转换 19-3
电平转换芯片 12-5
电气隔离 4-2
电气类型 17-4

电气特性 12-5
电气信号特性 9-3
电气栅格 17-2
电桥 1-2
电容 5-6
电容串并联 5-10
电容反馈式（电容三点式） 22-9
电容滤波 22-10
电容起动 3-10
电容式触摸屏 19-3
电容式传感器 1-2
电容式压力变送器 1-2
电容式液位探头 1-2
电容效应 4-1
电容元件 5-6
电容运行 3-10
电容值 3-5
电势 3-1
电枢 3-3
电枢磁场 4-4
电枢电流 3-4
电枢电压 18-1
电枢电阻 18-1
电枢反应 3-3
电枢控制 4-4
电枢直径 3-5
电梯控制 14-3
电通量 5-6
电网 14-3
电网电压 18-1
电网换流 4-5
电位 5-1
电位参考点 5-1
电位差（电势差） 5-1
电位计 3-5
电位降 5-1
电位升 5-1
电涡流式传感器 1-2
电压 5-1
电压（源）型逆变电路 4-5
电压比较器 21-7
电压表 5-1
电压表的负载效应 5-2
电压并联 22-7
电压传输特性 5-5

电压串联 22-7
电压电流关系 5-1
电压调整 3-3
电压反馈 18-1
电压反馈系数 18-1
电压放大模型 22-1
电压放大器 1-2
电压符号 5-9
电压负反馈 18-1
电压负反馈控制系统 18-1
电压跟随器 22-2
电压互感器 3-5
电压基值 3-5
电压极性 5-9
电压降 3-1
电压控制电流源 5-4
电压控制电压源 5-4
电压输出端 2-8
电压相量 5-8
电压源 5-4
电压源内阻 5-4
电压源逆变器 18-6
电液型 1-4
电源 5-1
电源变换 5-2
电源层 17-5
电源地 17-2
电源故障 6-2
电源描述符 15-4
电源模块 6-3
电源效率 22-1
电子 5-1
电子开关 12-4
电子控制单元 11-4
电子设计自动化 17-1
电子数据交换 14-6
电子锁 14-5
电阻 3-4
电阻的并联 5-2
电阻的串联 5-2
电阻电路 5-2
电阻分相 3-10
电阻式触摸屏 19-3
电阻网络 2-8
电阻-温度特性 1-3

电阻性阻抗 5-9
电阻元件 5-1
电阻值 3-4
吊车 3-4
掉电 2-2
迭代法 20-8
迭加驱动方式 19-3
叠加原理 5-4
叠片 3-3
叠绕组 3-3
顶层 17-5
顶层丝印层 17-5
顶隙 13-6
顶隙系教 13-6
订货号 6-3
定宽凸轮 13-5
定理 5-4
定量的 20-1
定律 5-4
定时控制 12-5
定时器 2-5
定时器方式 2-5
定时器类型 6-4
定时器中断 2-4
定位公差 13-10
定向公差 13-10
定性的 20-1
定值控制 11-3
定轴轮系 13-6
定子 3-3
动画连接 8-1
动力学系统 20-9
动量平衡 10-2
动能 3-2
动态标定 1-5
动态存储器 2-1
动态电路 5-11
动态电阻 5-11
动态分辨率选项 8-1
动态环流 18-3
动态平衡 3-2
动态数学模型 18-1
动态速降 18-2
动态特性 21-3
动态透明效果 17-1
动态显示 12-3

索引

动态线路 5-11
动态响应 1-2
动态性能 18-2
动态元件 5-6
动载荷 13-9
动作 6-5
动作电流 18-5
动作脚本程序 8-1
动作时间 6-2
抖动 12-3
都市办公大楼 14-1
读/写存储器 21-8
读卡机 14-5
读写命令 9-6
读周期 21-8
独立 I/O 2-2
独立电源 5-4
堵转电流 18-2
堵转实验 3-7
端点 15-3
端点电压 5-9
端点描述符 15-4
端点匹配 15-4
端电压 3-3
端盖 3-3
端环 3-7
端口地址 2-2
端口信号流向 17-2
端面凸轮 13-5
端子 3-5
端子符号 7-3
端子排 7-4
端子排编辑器 7-4
短路 5-1
短路保护 6-1
短路导纳参数矩阵 5-14
短路输出导纳 5-14
短路输入导纳 5-14
短路输入阻抗 5-14
短路转移导纳 5-14
断点 15-6
断开网络 15-5
断路器 3-8
断态 4-7
断态(阻断状态) 4-3
断态重复峰值电压 4-1

堆栈 2-2
对被选对象创建对象组合 17-3
对称的 5-14
对称电路 19-6
对称度 13-10
对称三相电路 5-12
对称网络 5-3
对称性 20-4
对分查表法 12-6
对话框 8-1
对角标准形 20-9
对角化 20-9
对流 3-8
对偶 20-9
对偶函数 21-2
对齐到栅格 17-3
对数阀 10-2
对数运算电路 22-2
对外等效 5-4
对象组合 17-3
对心从动件 13-5
对心曲柄滑块机构 13-4
多边形敷铜层 17-5
多变量模糊控制 12-9
多变量系统 12-9
多播标志域 15-5
多层印制板 17-5
多层印制电路板 17-5
多导线母线 7-3
多点传送 15-3
多点控制单元 14-3
多电平逆变电路 4-5
多发射极三极管 21-5
多发射结晶体管 22-4
多回路系统 20-2
多机通讯 2-6
多级泵 10-4
多级放大电路 22-4
多级计算机控制 12-1
多级行星轮系 13-5
多级与非门电路 21-3
多监视器系统 8-1
多晶硅 22-5
多路 A/D 转换器 2-8
多路电流源 22-6

多路数据分配器 21-4
多路数据选择器 21-4
多输入单输出 12-9
多速率采样 12-2
多跳路由 15-1
多线制 14-4
多项式 20-9
多楔带 13-7
多谐振荡器 21-7
多用户多媒体插座 14-6
多用户信息插座 14-6
多油楔滑动轴承 13-9
多元线性回归 16-6
多重化 4-5
多重逆变电路 4-5
多主广播模式串行总线 11-4
惰轮 13-6

E

额定电流 18-1
额定电压 3-3
额定负载 3-4
额定功率 3-4
额定容量 5-9
额定寿命 13-9
额定速降 18-1
额定转速 3-4
二次方程 10-3
二次击穿 4-1
二端电路 5-4
二端口耦合电感元件 5-10
二端口网络 5-14
二端口元件 5-14
二端网络 5-2
二端元件 5-2
二极管 21-3
二极管检波电路 1-2
二阶的 10-3
二阶惯性环节 12-8
二阶系统 20-3
二阶因子 20-5
二阶最佳 18-2
二进制 2-1
二进制编码的十进制数 21-1

二进制计数器 21-6
二进制结果位 6-4
二进制码 1-6
二—十进制译码器 21-4
二输入与门 21-3
二维草图与注释 7-1
二值数字逻辑 21-1

F

发出功率 3-5
发电机 3-3
发电机惯例 3-9
发光 22-3
发光二极管 2-2
发散 20-10
发散振荡 10-3
发射结电容 22-4
发送端 2-6
发送缓冲器 2-6
发送控制器 2-6
发送数据 2-6
发送中断标志 2-6
阀杆 10-2
阀门定位器 10-2
阀芯 10-2
法拉 5-6
法拉第电磁感应定律 5-6
法拉第效应式传感器 1-4
法兰 10-4
翻转 21-5
翻转触发器 21-6
翻转电路 2-2
反比于 3-5
反变换 20-3
反并联 4-4
反并联线路 18-3
反电动势 3-3
反函数 21-2
反激电路 4-5
反接/反向串联 5-10
反接制动 3-4
反馈 21-6
反馈差错控制 9-4
反馈电阻 2-8
反馈放大电路 22-7
反馈概念 22-7
反馈环节 20-1

反馈极性判断 22-7
反馈检测 18-1
反馈检测精度 18-1
反馈滤波 18-2
反馈通路 22-7
反馈网络 22-7
反馈校正 20-6
反馈信号 22-7
反馈组态 22-7
反码 2-1
反平行双曲柄机构 13-4
反推工程 19-5
反相 5-8
反相器 21-3
反相输入 5-5
反向饱和电流 22-4
反向二极管 4-1
反向二极晶闸管 4-1
反向恢复时间 4-1
反向击穿 4-1
反向截止状态 4-1
反向控制 4-5
反向偏置 22-4
反向偏置安全工作区 4-1
反向器 2-2
反向重复峰值电压 4-1
反向转动 12-4
反向阻断能力 4-1
反演规则 21-2
反正弦分布 16-2
反转 3-6
反组触发装置 18-3
反作用 12-7
范成法 13-6
方波 2-5
方波发生器 22-9
方差 16-2
方差分析 16-6
方焊盘 17-5
方式选择位 2-5
方向控制 12-4
防爆 3-7
防爆保护模拟量输出 6-7
防爆保护模拟量输入 6-7
防爆电动机 3-7
防尘盖 13-9

防尘盖轴承 13-9
防滴电动机 3-7
防焊膜 17-5
防火安全门 14-4
防火幕 14-4
防溅电动机 3-7
仿形法 13-6
仿真 15-6
仿真器 15-6
访问 2-1
访问时间 2-2
放大 8-1
放大电路 22-4
放大概念 22-4
放大机构 20-1
放大器 2-8
放大区 22-3
放大作用 22-4
放电 21-7
放热的 10-4
放元件 17-2
飞轮矩 3-2
非 21-2
非编码键盘 19-3
非标准齿轮 13-6
非电量检测仪 1-1
非独立(受控)电源 5-4
非负定 20-9
非接触式测温 1-7
非金属材料 13-2
非平面电路 5-3
非屏蔽双绞线 14-6
非稳态系统 18-1
非线性参数 12-6
非线性程度 10-2
非线性电感元件 5-6
非线性电路 5-1
非线性电容元件 5-6
非线性电阻 5-1
非线性电阻元件 5-1
非线性复杂系统 12-9
非线性函数的近似 10-3
非线性控制系统 20-1
非线性滤波器 20-10
非线性失真 22-4
非线性误差 1-1

非线性校正 19-4
非最小相位特性 10-3
非最小相位系统 20-4
废催化剂 10-4
分贝表示法 22-1
分辨率 2-8
分布电容 1-2
分布函数 16-2
分布密度 16-2
分布式 2-6
分布式 I/O 设备 6-2
分布式报警系统 8-1
分布式地址分配机制 15-5
分布式数据采集 19-2
分布式系统 2-6
分布式应用 8-1
分部积分法 5-13
分程控制 10-4
分布式控制系统 10-4
分段曲线拟合 19-4
分段线性插值 19-4
分解 3-6
分离点 20-4
分立电路 22-1
分流电路 5-2
分馏塔 10-4
分马力 3-10
分配率 21-2
分配线架 14-6
分片 15-4
分频 2-5
分频器 22-9
分散控制器 14-3
分时处理 20-8
分析 13-3
分析方法 22-7
分相 3-10
分压电路 5-2
分支 6-5
分支和合并 6-5
分支语句 8-1
分组法 16-6
风机盘管 14-3
风扇 3-4
风扇冷却 3-7

风阻损耗 3-7
封锁延时 18-3
封装 17-7
峰-峰值 4-5
峰值 3-5
峰值检波器 22-1
峰值时间 20-3
伏 3-1
伏安特性 22-3
伏特 5-1
服务发现 15-4
服务器 8-1
浮点数 2-1
浮置 19-6
符号 5-10
符号编辑器 6-4
符号编译器 7-3
符号数 21-1
幅频特性 22-4
幅频图 20-5
幅移键控法 12-5
幅值 22-9
幅值/振幅 5-8
幅值控制 18-1
幅值条件 20-6
幅值相量 5-8
幅值裕度 22-4
辐射 3-8
辅助电源 4-3
辅助进位标志 2-2
辅助绕组 3-10
负边沿 21-5
负电荷 5-1
负电源 5-5
负定 20-9
负反馈 22-7
负反馈电路的自激振荡 22-7
负号 21-1
负荷 14-3
负逻辑 21-3
负脉冲 2-5
负实部 20-3
负跳沿跳转 6-4
负相序 3-5
负载 5-4

负载电流 18-1
负载电路 4-1
负载换流 4-5
负载能力 2-2
负载情况下 3-3
负载特性 3-3
负载线 22-4
负载线分析 22-4
负载效应 22-1
负载转矩 3-4
附加电流源 5-13
附加电压源 5-13
附加端子 6-3
附加方程 5-3
附加噪声 22-1
附属设备 4-3
复功率 5-9
复合铰链 13-3
复合数字滤波 12-6
复合校正 20-6
复合轴承材料 13-9
复励发电机 3-3
复平面 20-4
复数 10-3
复数根 20-3
复数增益 22-1
复位 2-2
复位设备 15-5
复位状态 2-4
复现 20-8
复相 20-8
复用 2-2
复杂对象 8-1
复杂可编程逻辑器件 21-9
复杂描述符 15-4
复制端子特性 7-4
复阻抗 5-9
副边 3-5
副边漏电抗 3-5
副调节器 10-4
副回路 10-4
傅立叶变换 10-3
覆盖窗口 8-1

G

干扰 20-1
干扰信号 21-3
干涉式温度传感器 1-3
干式变压器 3-5
感抗 5-9
感纳 5-9
感性 5-9
感性阻抗 5-9
感应 3-6
感应电动机 3-6
感应电势 1-4
感应同步器 1-6
感应同步式 1-6
刚度 13-2
刚性冲击 13-5
刚性联轴器 13-8
高电平触发 2-2
高副 13-3
高级智能 19-1
高阶保持器 12-8
高阶系统 20-3
高频 20-5
高频噪声 19-4
高速数据通道 11-2
高通 19-2
高通滤波器 22-9
高温试验 19-6
高选器 10-4
高压边 3-5
高压集成电路 4-1
高压绕组 3-5
高阻态 2-2
格雷码 21-1
格罗布斯准则 16-2
隔离放大器 19-2
隔离技术 19-6
隔离开关 3-8
个人计算机 2-1
个人识别号 14-5
给定电压 18-1
给定积分器 18-6
给定量 20-1
给定滤波 18-2
给定值 12-7
给排水设备 14-3
给排水系统 14-3
根轨迹 20-2

根集线器 19-3
跟随性能 18-2
跟踪误差 20-7
工厂操作员 8-1
工程计算 20-3
工程软件 19-5
工程设计自动化 17-1
工程师站 11-2
工程塑料 13-2
工具栏 7-1
工具面板 8-1
工具软件 19-5
工控机 11-1
工位号 8-1
工业过程控制 12-1
工业控制机 12-10
工业控制组态软件 12-6
工业生产对象 12-1
工业以太网 9-1
工业自动化系统 12-1
工艺流程 12-5
工艺流程图 10-1
工作标准 15-1
工作齿廓 13-6
工作存储器 6-2
工作点 21-3
工作点稳定 22-4
工作电压 5-4
工作方式 2-5
工作寄存器 2-1
工作流量特性 10-2
工作行程 13-4
工作原理图 6-1
公差带 13-10
公差带代号 13-10
公差等级 13-10
公差原则 13-10
公法线长度 13-6
公共端 5-5
公共广播 14-4
公共广播系统 14-4
公共规范 15-3
公制单位 17-2
功 3-1
功耗 21-3
功角特性 3-9

功率 5-1
功率表 3-5
功率传输 3-5
功率放大电路 22-8
功率集成电路 4-1
功率模块 4-6
功率三角形 5-9
功率因数 5-9
功率因数的提高 5-9
功率因数角 5-9
功率因数校正 4-6
功率增益 22-1
功能表 21-4
功能键 12-3
功能块 6-2
功能块库 6-4
功能码 9-6
功能实体 15-5
供电电压 4-8
汞弧整流器 4-1
共轭复数 3-5
共轭匹配 5-11
共轭虚根 20-3
共基组态 22-4
共集组态 22-4
共价键 22-3
共模放大倍数 22-6
共模信号 22-6
共模抑制比 22-6
共设截止频率 22-4
共射组态 22-4
共享数据块 6-4
共振式光导纤维振动传感器 1-5
钩头楔键 13-8
构件 13-1
固定床操作 10-4
固定电阻 5-11
固定流化床 10-4
固态传感器 1-4
固态继电器 4-1
固态图像传感器 1-4
固态温度传感器 1-4
固态整流器 3-9
固体激光器 1-5
固有频率 1-6

固有频率/自然频率 5-11
关断 4-6
关断电流 4-1
关断过程 4-1
关断过电压 4-1
关联参考方向(无源符号约定) 5-1
关联端子 7-4
关联要素 13-10
关中断 2-5
管道仪表流程图 10-1
管脚 21-4
管口 10-2
管理信息系统 12-1
管理自动化 14-1
管式加热炉 10-4
管线 10-2
管线调和 10-4
惯性 3-2
惯性环节 20-2
惯性系数 18-1
罐的呼吸 10-4
光/电转换器 14-4
光标位置 17-5
光导纤维 1-5
光导纤维 FF 型传感器 1-5
光点矩 19-3
光电池 1-4
光电传感器 1-4
光电导探测器 1-4
光电二极管 22-3
光电隔离技术 12-4
光电检测器 1-5
光电脉冲发生器 1-5
光电耦合器 22-4
光电效应 1-4
光电元件 1-4
光电转速计 1-4
光隔离 6-3
光控晶闸管 4-1
光敏传感器 1-4
光敏电阻 1-4
光敏二极管 1-4
光敏区 1-4
光盘 2-1

光谱特性 1-4
光驱 2-1
光纤传感器 1-5
光纤磁场电流传感器 1-5
光纤到家庭 14-6
光纤到桌面 14-6
光纤电缆 6-6
光纤互连 14-6
光纤水声传感器 1-5
光纤同轴电缆混合系统 14-6
光纤转速传感器 1-5
光源 1-5
光栅传感器 1-6
光柱显示器 12-3
广播 15-3
广义对象 12-8
广义结点 5-3
广义网孔 5-3
广域网 9-7
规则采样法 4-6
滚齿机 13-6
滚动体 13-9
滚动轴承 13-9
滚针 13-9
滚针轴承 13-9
滚柱离合器 13-8
滚子从动件 13-5
国际标准化组织 9-1
国际单位制 5-1
国际电工委员会 12-5
过补偿 18-1
过采样技术 19-2
过程 19-5
过程动态学 10-1
过程静态增益 10-3
过程开发模型 19-5
过程控制 10-1
过程模型 10-2
过程设备 10-4
过程现场总线 12-5
过程映像输入寄存器 6-2
过导孔 17-5
过电流 18-1
过电流保护 4-1
过电压 18-1

过电压保护 4-1
过渡过程 3-2
过渡配合 13-10
过孔 17-5
过孔类型选择 17-5
过励磁 3-9
过流保护 6-1
过热保护 22-10
过热器 10-4
过热蒸汽 10-4
过盈 13-10
过盈配合 13-10
过载 3-7
过载继电器 3-8
过阻尼 12-7
过阻尼系统 10-3

H
函数发生器 18-6
函数关系 16-6
函数随机误差 16-3
函数误差 16-3
函数系统误差 16-3
汉明码和汉明距离 9-4
焊接面 17-5
焊盘 17-5
焊盘孔 17-5
航空无线电公司分配法 19-6
毫亨 5-6
耗尽层 22-3
合成标准不确定度 16-4
合金钢 13-2
赫尔维兹判据 20-3
赫兹 3-1
亨利 5-6
恒等式 21-2
恒功率方式 3-4
恒功率负载 3-4
恒流区 22-3
恒流源 22-6
恒压恒频 4-8
恒转矩方式 3-4
恒转矩负载 3-4
横截面积 10-2
横向效应 1-2
红外线式触摸屏 19-3

红外线式温度传感器 1-3
红外线探测器 14-5
宏 15-6
宏单元 21-9
后台调试模式 19-5
后缀设置 7-1
厚度测量 1-7
呼吸阀 10-4
互补电路 22-8
互补金属氧化物半导体门电路 21-1
互导 5-3
互感 5-10
互感电抗 5-10
互感式传感器 1-2
互换性 13-10
互锁 6-1
互易定理 5-4
互易条件 5-4
互易网络 5-4
互阻 5-3
互阻放大模型 22-1
花键 13-8
滑差 3-6
滑动率 13-7
滑动平均 19-4
滑动平均滤波 12-6
滑动轴承 13-9
滑动轴承 3-7
滑环 3-3
滑键 13-8
滑块 13-4
滑块的导路 13-4
滑块联轴器 13-8
滑轮 3-2
化简 21-2
化学工程 10-1
化学热处理 13-2
画面 8-1
环节 20-1
环境参数设置 17-1
环境温度 1-3
环流 4-2
环流电抗器 18-3
环路增益 22-7
环形计数器 21-6

缓冲 2-2
缓冲电路 4-3
缓冲电容 4-3
缓冲器 2-1
换流 4-5
换路 5-7
换路开关 18-4
换能器 18-5
换向 3-3
换向过电压 4-1
换向极 3-4
换向片 3-3
换向片数 3-3
换向器 3-3
换向重叠角 4-2
换行 2-1
灰铸铁 13-2
恢复布线 17-6
恢复时间 18-2
恢复特性 4-1
恢复现场 2-4
辉光数码管 21-4
回差电压 21-7
回车 2-1
回归方程 16-6
回归分析 16-6
回归直线 16-6
回火 13-2
回馈制动 3-8
回流 10-4
回路 5-3
回路编译器 7-3
回路电流 5-3
回路电流法 5-3
回路分析法 5-3
回路配置 7-3
回路元素 7-3
汇编 2-3
汇编程序 2-3
汇编错误码 2-3
汇编符号集 2-3
汇编语言 2-3
汇合点 20-4
会话层 9-5
会议电视系统 14-6
绘图仪 2-1

绘图应用 17-2
混合电 17-5
混合纠错 12-5
混合轮系 13-6
混合信号仿真 17-2
混频器 22-9
火花 3-3
火警系统 14-4
火炬 10-4
火灾管理系统 14-4
火灾红外探测器 14-4
火灾控制器 14-4
或 21-2
或非 21-2
或然误差 16-2
霍尔系数 1-4
霍尔效应 1-4
霍尔压力变送器 1-4
霍尔元件 1-4
霍耳电流变换器 18-1
霍尼韦尔 14-2

J

击穿 4-3
击穿电流 4-3
击穿电压 22-3
击穿时间 18-3
击穿特性 4-3
击键 2-7
机电能量转换 3-1
机电时间常数 18-1
机电时间常数 3-4
机构 13-1
机构运动简图 13-3
机架 13-3
机壳 3-7
机器 13-1
机器语言 2-3
机器周期 2-2
机械 13-1
机械的 13-1
机械动力装置 13-1
机械功率 3-4
机械角度 3-6
机械开关 21-5
机械离合器 13-8
机械量检测 1-7

机械能 3-2
机械试验 19-6
机械特性 3-3
机械学 13-1
机械原理 13-1
机械装置 13-1
奇函数 5-7
奇偶 2-2
奇偶发生/检验器 21-2
奇偶校验 12-5
奇偶校验位 2-6
奇数 20-9
奇校验 21-2
积分饱和 12-7
积分补偿 12-7
积分调节器 18-1
积分环节 20-2
积分控制 18-1
积分时间 12-7
积分性质 5-13
积分运算电路 22-2
积复励发电机 3-3
基本 RS 触发器 21-5
基本尺寸 13-10
基本额定寿命 13-9
基本方程 22-7
基本偏差 13-10
基本项目 6-4
基波 3-6
基波频率 3-6
基波因数 4-2
基础导线 7-3
基尔霍夫电流定律 5-1
基尔霍夫电流方程 3-1
基尔霍夫电压定律 5-1
基尔霍夫电压方程 3-1
基尔霍夫定律 5-1
基孔制配合 13-10
基数 2-1
基于参考的标记 7-1
基于信息的协议 11-4
基圆 13-5
基圆齿厚 13-6
基圆直径 13-6
基值 3-5
基址加变址寻址 2-3

基轴制配合 13-10
基准电流 2-8
基准电压 2-8
基准孔 13-10
基准宽度 13-7
基准源 22-10
基准直径 13-7
基准制 13-10
基准轴 13-10
畸变 20-10
畸变功率 4-2
激光扫描测量装置 1-5
激励 5-4
激励表 21-6
激励电压 1-2
级联放大器 22-1
级数求和 20-8
即插即用 19-3
极差法 16-2
极点 20-4
极点配置 20-9
极间电容 22-3
极距 3-6
极数 3-3
极限尺寸 13-10
极限环 20-10
极限啮合点 13-6
极限偏差 13-10
极限误差 16-2
极限误差的合成 16-3
极性标注 3-5
极坐标图 20-5
极坐标形式 5-8
急回特性 13-4
急回特性机构 13-4
急回运动 13-4
急停按钮 6-1
集成 2-2
集成测试 19-5
集成传感器 19-2
集成电路 22-1
集成电路特点 22-6
集成电压比较器 22-9
集成功放 22-8
集成红外CCD固态图像传感器 1-5

集成开发环境 8-1
集成门极换流晶闸管 4-1
集成温度传感器 1-3
集成元件库 17-4
集成运算放大电路 22-6
集电极开路 21-3
集电结电容 22-4
集散控制系统 11-1
集散型控制系统 14-2
集中参数 5-2
集中参数电路 5-2
集中参数元件 5-2
集中管理 11-1
集中式数据采集 19-2
集总布线区 21-9
几何平均 20-6
计量泵 10-4
计数长度 2-5
计数法 12-4
计数范围 6-4
计数模式 21-6
计数器 2-5
计数器方式 2-5
计数器和比较指令 6-4
计数器中断 2-4
计算查表法 12-6
计算机辅助测试 17-1
计算机辅助电路分析 5-11
计算机辅助软件工程 19-5
计算机辅助设计 17-1
计算机辅助制图 17-1
计算机辅助制造 17-1
计算机集成制造系统 12-1
计算机技术 19-1
计算机监督控制 12-1
计算机控制技术 12-1
计算机控制系统 12-1
计算数据比较法 16-2
记录 8-1
记录仪表 12-3
记忆元件 5-6
技术支持 8-1
技术指标 20-6

继承性 19-5
继电器 3-8
继电器触点 6-1
继电器特性 20-10
继电器线圈 7-3
继电效应 1-3
寄存器 2-1
寄存器间接寻址 2-3
寄存器寻址 2-3
寄生电容 1-2
寄生直流电势 1-4
加标注 17-5
加法电路 5-5
加法器 2-1
加法摄动 20-7
加密 15-2
加氢精制 10-4
加氢裂化 10-4
加权 19-4
加权平均滤波 12-6
加权算数平均值 16-2
加热炉 10-4
加热盘管 10-4
加热器 10-4
加入网络 15-5
加数 21-4
加速 3-2
加速度 20-3
加速度传感器 1-2
夹断电压 22-5
夹壳联轴器 13-8
甲类放大 22-8
甲烷化 10-4
假分式 5-13
尖端从动件 13-5
间接标识符 8-1
间接电流控制 4-8
间接寻址 15-3
间接直流变换电路 4-8
间隙 13-10
间隙配合 13-10
间歇运动的连杆机构 13-4
兼容性指导 8-1
监督式控制 11-2
监控程序 19-5

监控管理程序 12-1
监控设备 14-4
监控室 14-4
监控台 14-4
监控系统 14-4
监视器 6-2
减法器 2-1
减数 21-4
减速 3-2
减速比 13-7
减压塔 10-4
剪切频率 20-5
剪贴板 17-3
检测 2-2
检测电路 4-1
检测技术 1-1
检测元件 12-1
检查工具 17-2
检索 8-1
简单比较器 22-9
简单描述符 15-4
简化的 OSI 模型 9-5
简化功能设备 15-3
简图 13-3
简谐运动规律 13-5
简写符号 13-7
建筑群配线架 14-6
建筑物布线系统 14-6
建筑物配线架 14-6
建筑物智能系统集成 14-6
渐近稳定 20-3
渐近线 20-4
渐开线齿轮 13-6
渐开线函数 13-6
渐开线花键 13-8
鉴幅器 18-1
键 13-8
键槽 13-8
键码 2-7
键盘 2-2
键盘接口 2-7
键盘接口技术 2-7
键盘扫描 2-7
键盘事件 2-7
键扫描程序 2-7

江森自控 14-2
降压变压器 3-5
降压斩波器 4-3
交叉开发 19-5
交互参考 7-1
交互式布线 17-5
交互式电视 14-6
交换率 21-2
变频电路 4-4
交-交变频器 18-6
交流 18-1
交流并励电动机 18-6
交流潮流 18-6
交流串励电动机 18-6
交流电动机 12-4
交流电机 18-1
交流电力电子开关 4-4
交流电力控制 18-5
交流电流 3-1
交流电气传动 18-1
交流电源 5-2
交流调功电路 4-4
交流调速 18-5
交流调压电路 4-4
交流分量 18-6
交流负载线 22-4
交流互感器 18-1
交流换向器电动机 18-6
交流继电器 12-4
交流鼠笼电机 4-5
交流通路 22-4
交越失真 22-8
交-直-交变频器 18-6
交-直流变流器 18-6
交直流电动机 3-10
交轴 3-9
焦点 20-10
焦耳 5-1
角度标注 17-5
角接触推力轴承 13-9
角接触轴承 13-9
角频率 5-8
矫顽力 3-1
脚本 8-1
脚本触发器 8-1
脚本语言 8-1

索引

铰链四杆机构 13-4
校正 1-1
校正措施 22-4
校正动作 10-2
校正函数 19-4
阶码 2-1
阶跃函数 5-7
阶跃输入 18-2
阶跃响应 5-7
阶跃信号 20-3
接触电势 1-3
接触器 3-8
接触式测温 1-7
接触线、啮合线 13-6
接地故障检测器 6-3
接地开关 3-8
接口 2-2
接口标准 9-3
接口电路 2-2
接收电路 19-6
接收端 2-6
接收缓冲器 2-6
接收机同步 15-5
接收控制器 2-6
接收数据 2-6
接收中断标志 2-6
接受授权协议 17-1
接通延迟定时器 6-4
接线端块 6-1
接线端子 7-2
接线连接点 7-2
节点 5-3
节点描述符 15-4
节距 3-6
节宽 13-7
结点电压 5-3
结点电压法 5-3
结点分析法 5-3
结构 22-3
结构分析 13-3
结构化布线系统 14-6
结构化文本 6-2
结垢 10-4
结合率 21-2
结束分隔符 11-4
结型 22-3

截止电流特性 18-1
截止阀 10-4
截止频率 18-1
截止区 22-3
解除被选的对象 17-3
解码网络 12-2
解耦控制 10-4
解析法 20-7
介电常数 5-7
介质访问控制 6-6
介质接口连接器 14-6
借位 21-4
金属箔双绞电缆 14-6
金属材料 13-2
金属导轨 6-2
金属氧化物绝缘栅场效应管 22-5
紧边拉力 13-7
紧耦合 3-5
进位 21-4
进位标志 2-2
禁止中断 2-4
经典控制理论 20-1
经济性 19-6
经济性能 20-6
晶片 22-9
晶体管 5-5
晶体管-晶体管逻辑 21-1
晶体管输出 6-3
晶闸管 4-1
晶闸管的并联 4-1
晶闸管的串联 4-1
晶闸管控制电抗器 4-4
晶闸管投切电容器 4-4
晶闸管相控变换器 4-2
晶振 2-2
晶振频率 crystal 2-2
精度 16-1
精度估计 16-5
精馏塔 10-4
精密度 16-3
精密机械 13-1
精确度 16-1
径向滑动轴承 13-9
径向间隙 13-6
径向载荷系数 13-9

径向-止推滑动轴承 13-9
竞争冒险 21-2
静差率 18-1
静电感应晶体管 4-1
静电感应晶闸管 4-1
静态变量 6-4
静态标定 1-5
静态存储器 2-1
静态工作点 22-4
静态环流 18-3
静态驱动方式 19-3
静态特性 4-1
静态误差 12-7
静态误差系数 20-6
静态显示 12-3
静态性能 20-4
静载荷 13-9
静止磁场 3-6
静止无功补偿器 4-7
镜像 17-5
镜像电流源 22-6
纠错编码 12-5
局部操作网络 12-5
局部实际尺寸 13-10
局部自由度 13-3
局域网 9-1
菊花链 11-4
矩角特性 3-9
矩形波发生器 22-9
矩形花键 13-8
矩形脉冲 10-3
矩形填充 17-2
矩阵 20-3
矩阵键盘 12-2
矩阵式变频电路 4-5
矩阵式变频器 4-4
锯齿波调制 4-6
锯齿剥发生器 22-9
聚合装置 12-1
卷积定理 20-8
决策支持系统 14-6
绝对地址 6-4
绝对精度 19-2
绝对偏差 19-4
绝对湿度 1-4
绝对误差 16-1

绝对寻址 6-4
绝对值变换器 18-6
绝缘 3-3
绝缘等级 3-5
绝缘体 5-7
绝缘栅双极晶体管 4-1
绝缘栅型 22-5
均等分配法 19-6
均方根 3-6
均流 4-1
均压 4-1
均匀分布 16-2

K

卡口式光纤连接器 14-6
卡诺图 21-2
开槽 3-7
开度 10-2
开发 19-5
开发系统 8-1
开放的全球标准 15-2
开放结构控制器 11-4
开放式控制系统 12-1
开放系统互连参考模型 9-1
开放延时 18-3
开关 5-7
开关电源 4-3
开关高频二极管的性能 22-3
开关关断特性 4-3
开关柜 6-1
开关函数 5-7
开关量 12-2
开关时间 21-3
开关瞬态过程 4-3
开关损耗 4-3
开关特性 4-3
开关通态特性 4-3
开关型直流电源 22-10
开关噪声 4-3
开环 18-1
开环传递函数 20-2
开环电压增益 5-5
开环放大倍数 18-1
开环极点 20-4
开环控制系统 20-1

开环增益 20-4
开口销 13-8
开路 5-1
开路输出导纳 5-14
开路输出阻抗 5-14
开路输入阻抗 5-14
开路转移阻抗 5-14
开路阻抗参数矩阵 5-14
开启电压 4-1
开始啮合点 13-6
开式链 13-3
开通 4-3
开通过程 4-1
开尾圆锥销 13-8
开源操作系统 15-1
开中断 2-5
看门狗 19-6
抗干扰 19-6
抗干扰技术 19-6
抗积分饱和 12-7
抗拉强度 13-2
抗扰性能 18-1
可编程 21-8
可编程接口 2-2
可编程控制器 6-2
可编程逻辑控制 6-2
可编程逻辑控制器 12-1
可编程序控制器 11-1
可编程增益放大器 12-6
可编程阵列逻辑 21-8
可变电容器 18-5
可变电阻 5-9
可擦可编 21-8
可擦写的只读存储器 2-1
可调比 10-2
可调变压器 18-5
可调电抗器 18-5
可调电阻器 18-5
可调恒速电动机 18-5
可观测的 20-9
可靠率 19-6
可靠性 10-4
可控的 20-9
可控硅 3-8
可控硅接口技术 12-4
可控硅整流器 4-1

可控整流 4-2
可能状态 21-5
可逆 PWM 变换器 18-4
可逆计数器 21-6
可逆线路 18-3
可逆循环 4-5
可配置逻辑块 21-9
可视背景栅格 17-2
可视化对象 8-1
可视化元件 8-1
可寻址远程传感器数据通路 14-2
可寻址远程传感器数据通路 12-5
可用的库 17-2
可用性 19-6
可重复触发 21-7
可重用软件 19-5
客户端 8-1
空调系统 14-3
空回行程 13-4
空间基波分量 3-9
空间连杆机构 13-4
空间凸轮 13-5
空气处理机 14-3
空气断路器 18-3
空气开关 18-3
空气-燃料混合物 10-4
空闲位 2-6
空心变压器 5-10
空心线圈 5-10
空心销轴 13-8
空穴 22-3
空载 3-3
空载电流 18-1
空载实验 3-7
孔 13-10
孔板 10-4
控件 8-1
控制变量 10-5
控制传输 19-3
控制电路 4-1
控制对象 18-2
控制阀 12-7
控制方案 12-10
控制方式 12-7

控制规律 12-8
控制规则 12-9
控制回路 12-7
控制局域网 14-2
控制矩阵 20-9
控制理论 20-1
控制模式 18-6
控制模式 4-6
控制器 10-2
控制器局域网络 12-5
控制算法 11-3
控制位 2-4
控制系统 10-1
控制信号 2-2
控制性能指标 10-3
控制质量 10-3
控制转移 2-3
控制总线 2-2
库仑 5-1
库选项 17-7
跨区域寄存器间接寻址 6-4
块调用命令 6-4
块接口 6-4
块结构 6-4
块输入 6-4
块图标 6-4
块校验 9-4
快恢复二极管 4-1
快恢复外延二极管 4-1
快开阀 10-2
快闪存储器 21-8
快速晶闸管 4-1
快速熔断器 4-1
快速移动元件 7-3
快速以太网 9-7
宽 V 带 13-7
宽度系列（滚动轴承）13-9
框图 3-8
扩散电容 22-3
扩展 21-4
扩展模块 6-3
扩展帧格式 11-4

L

拉普拉斯变换 5-13

拉电流 21-3
拉普拉斯变换对 5-13
拉普拉斯反变换 5-13
拉普拉斯逆变换 12-7
来自低位进位 21-4
蓝牙 15-2
浪涌电流 4-1
浪涌电压 4-1
劳斯稳定判据 20-3
泪滴焊盘 17-5
累加器 2-1
楞次定律 5-6
冷端补偿器 1-3
冷端温度 1-3
离合器 13-8
离散控制系统 12-7
离散时间系统 12-2
离散时间信号 12-1
离散系统 12-8
离散信号 21-1
离线 12-10
离心拉力 13-7
离心离合器 13-8
离心力 3-4
李亚普诺夫方程 20-9
李亚普诺夫稳定性判据 20-9
理论啮合线 13-6
理想变压器 5-10
理想电路元件 5-2
理想二极管模型 22-3
理想空载 18-1
理想空载转速 18-1
理想流量特性 10-2
理想模型 22-3
理想要素 13-10
理想运放 22-6
理想运算放大器 5-5
力敏效应 1-4
力敏元件 1-2
力学量测量 1-7
历史报警 8-1
历史趋势 8-1
立即寻址 2-3
励磁变阻器 3-4
励磁磁通 18-1

励磁电抗 3-7
励磁电流 3-3
励磁电流调节器 18-2
励磁绕组 3-3
励磁线圈 3-3
连杆 13-4
连杆曲线 13-4
连击 12-3
连接点符号 17-2
连接方式 22-4
连锁保护 18-3
连通图 5-3
连续的 5-7
连续过程 10-1
连续控制 6-7
连续控制系统 12-7
连续量 2-8
连续时间系统 12-2
连续时间信号 12-1
连续信号 21-1
连支 5-3
联锁操作 14-3
联轴器 13-8
联组 V 带 13-7
链接 15-6
链接元件 7-3
链路质量指示 15-3
链条联轴器 13-8
链系 13-3
两表法 5-12
量程自动转换 12-6
量化 19-2
量化误差 21-10
列向量 20-9
列选择线 21-8
裂解炉 10-4
临界 21-3
临界比例度法 12-7
临界放大倍数 18-1
临界速度 10-4
临界稳定 20-3
临界压力 1-2
临界增益 20-7
临界阻尼 20-3
临界阻尼的 10-3
临时 2-2

灵敏度 1-1
灵敏度漂移 10-2
零初始条件 5-13
零点 20-4
零点残余电压 1-2
零点和极点 12-8
零点漂移 22-4
零电流 4-7
零电流检测器 18-3
零电流开关准谐振电路 4-7
零电流转换 PWM 电路 4-7
零电位点 5-1
零电压 4-7
零电压开关多谐振电路 4-7
零电压开关准谐振电路 4-7
零电压转换 PWM 电路 4-7
零件 13-1
零件目录 7-3
零阶保持器 12-8
零开关 4-7
零偏置 4-1
零输入响应 5-7
零位误差 19-4
零线 13-10
零转换 4-7
零状态 5-7
零状态响应 5-7
领先 3-5
令牌传递 9-7
浏览库元件 17-2
流程图 12-7
流程图 22-1
流化催化裂化 10-4
流量 12-7
流量变送器 1-4
流量传感器 14-2
流量计 1-7
流量检测 1-7
流量开关 14-3
流量系数 10-2
流速 10-2

流体 10-4
流体力学 10-2
留数 20-8
留数法 5-13
馏分 10-4
聋哑传感器 19-1
楼层配线架 14-6
楼宇自动化系统 14-1
漏磁通 3-5
漏电抗 3-7
漏电流 6-1
漏感 4-2
漏极 22-5
漏气 10-4
炉膛 10-4
鲁棒性 20-7
滤波 19-2
滤波电容 22-10
路径 5-7
路径的发现和选择 15-5
路径期满 15-5
路由错误报告 15-5
路由发现 15-5
路由器 15-3
路由协议 15-5
路由修复 15-5
铝合金 13-2
轮廓算术平均偏差 13-10
轮廓中线 13-10
轮廓最大高度 13-10
轮胎式联轴器 13-8
罗曼诺夫斯基准则 16-2
逻辑变量 21-2
逻辑表达式 21-2
逻辑操作结果 6-4
逻辑常量 21-2
逻辑单元阵列 21-9
逻辑电路 17-2
逻辑电平 2-2
逻辑仿真 17-2
逻辑符号 21-2
逻辑函数 21-2
逻辑控制 18-3
逻辑控制器 18-3
逻辑设计 17-2
逻辑条件脚本程序 8-1

逻辑图 17-2
逻辑运算 2-3
螺旋齿轮 13-6
螺旋模型 19-5
螺旋线 20-10
落后 3-5

M

马力 3-10
码盘 1-6
码制转换器 21-4
埋孔 17-5
脉冲 2-2
脉冲传递函数 12-8
脉冲封锁 18-3
脉冲幅度调制 18-6
脉冲干扰 19-4
脉冲后沿 4-6
脉冲宽度 4-6
脉冲宽度幅度变换器 4-6
脉冲宽度调制 12-4
脉冲频率调制 18-1
脉冲平均电路 4-3
脉冲前沿 4-6
脉冲式数字传感器 1-6
脉冲输入 14-3
脉冲限幅 18-6
脉冲响应 20-2
脉冲信号 20-3
脉冲序列 20-2
脉冲载波 4-3
脉冲重复频率 21-1
脉动系数 22-10
脉动因数 18-6
脉宽调制 4-3
脉振磁通势 3-10
满量程 10-2
满载 3-3
曼彻斯特编码电流调制 11-4
盲孔 17-5
梅花形弹性联轴器 13-8
梅逊公式 20-2
每分钟加仑 10-2
每分钟转数 3-6
每极磁通 3-6
门极 4-1

门极触发信号 18-1
门极关断开关 4-1
门极可关断晶闸管 18-4
门槛电压 4-1
门禁 14-5
门阵列 19-1
密封圈 13-9
密封圈轴承 13-9
密码用户 8-1
面板布局符号 7-4
面板元件 7-4
面对面配置(滚动轴承) 13-9
面结型二极管的物理机理 22-3
面轮廓度 13-10
面向对象 19-5
面向控制的编程语言 11-3
描述法 12-3
描述函数 20-10
描述函数法 20-10
描述系统 17-1
灭弧元件 6-1
敏感元件 1-1
敏三极管 1-4
铭牌 3-3
铭牌示意图 7-4
命令帧 15-5
模 21-6
模/数 14-3
模板 8-1
模糊关系 12-9
模糊控制 12-9
模糊控制算法 12-9
模糊控制系统 12-1
模糊数学模型 12-9
模糊条件语句 12-9
模糊语言 12-9
模块背板 6-3
模块化
模块化闭环控制 6-7
模块化楼宇控制器 14-6
模块化设备控制器 14-6
模块化设计 19-5
模块式结构 12-5

模拟的 20-8
模拟电流信号 2-8
模拟电压信号 2-8
模拟开关 21-3
模拟控制系统 12-1
模拟量 2-8
模拟量报警 8-1
模拟量输出 14-3
模拟量输出模块 6-3
模拟量输出通道 12-2
模拟量输入 14-3
模拟量输入模块 6-3
模拟量数入通道 12-2
模拟量信号 6-3
模拟滤波 12-6
模拟器模块 6-3
模拟信号 21-1
模式切换 11-3
模数转换 2-2
模数转换器 2-8
模型误差 20-7
膜盒 1-7
摩擦轮传动 13-7
摩擦式离合器 13-8
摩擦损耗 3-4
摩根定理 21-2
磨合性 13-9
磨损度 13-9
默认空白图纸尺寸 17-2
目标机 19-5
目标箭头 7-2
目标信号 7-2
目的 2-1
目的操作数 2-3
目的地址域 15-5
目的文件夹 17-1
目录查找 7-4

N

内部存储器 2-2
内部地址总线 2-2
内部电压降 18-1
内部控制总线 2-2
内部连线 21-9
内部数据总线 2-2
内部中断 2-4
内部总线 2-2

内存条 2-1
内径 3-7
内圈 13-9
内置通信端口 6-6
内置指针 6-4
乃奎斯特图 20-5
乃奎斯特稳定判据 20-5
能观性 20-9
能观性矩阵 20-9
能耗制动 3-4
能控标准形 20-9
能控性 20-9
能控性矩阵 20-9
能量 5-1
能量检测 15-3
能量平衡 10-2
能量守恒定律 5-9
逆变 4-5
逆变变压器 18-1
逆变电源 18-1
逆变器 3-8
逆导晶闸管 18-1
逆时针 3-3
逆时针方向 5-9
逆时针旋转 8-1
逆向电流 4-1
啮合点 13-6
啮合轨迹 13-6
啮合角 13-6
啮合平面 13-6
啮合线 13-6
暖启动 6-4
暖通空调 14-3
诺顿等效电路 5-4
诺顿定理 5-4

O

欧拉公式 5-8
欧姆 5-1
欧姆表 3-1
欧姆定律 5-1
偶函数 5-7
偶极子 20-6
偶然误差与系统误差的合成 16-3
偶数 20-9
偶校验 21-2

耦合 3-5
耦合电感元件 5-10
耦合电容 22-4
耦合通道 19-6
耦合系数 5-10
耦合线圈 5-10

P

排列与对齐 17-3
排气 10-4
盘形凸轮 13-5
旁路电容 22-4
抛物线 20-3
配电变压器 3-5
配电系统 14-3
配方管理器 8-1
配合 13-10
配合表面 13-10
配合公差 13-10
配合公差带 13-10
喷洒灭火系统 14-4
批量传输 19-3
皮法 5-6
疲劳极限 13-2
匹配 5-4
片上系统 15-6
片选 2-2
片选信号 2-2
偏差 10-2
偏差报警 8-1
偏差变量 10-3
偏差电压 18-1
偏心率 13-9
偏移量 2-1
偏移误差 19-2
偏移正交相移键控 15-2
偏振调制 1-5
偏置从动件 13-5
偏置电流 22-4
偏置曲柄滑块机构 13-4
漂移电流 4-1
频带 5-11
频带宽度 20-5
频率 5-8
频率计 21-6
频率特性 22-4
频率响应 20-2

索　引

频率选择电路 5-11
频谱估计 19-4
频移键控法 12-5
频域函数 5-13
频域特性 20-5
品质因数 5-11
平板电脑 8-1
平波电抗器 18-1
平带传动 13-7
平底从动件 13-5
平方根标度变换 8-1
平衡与不平衡传输线路 9-3
平键 13-8
平均功率 5-9
平均故障间隔时间 19-6
平均速度 12-4
平均通态电流 4-1
平均无故障时间 12-1
平均误差 16-2
平均修复时间 19-6
平均值 5-9
平面电路 5-3
平面度 13-10
平面副 13-3
平面机构 13-3
平面铰链四杆机构 13-4
平面连杆机构 13-4
平行度 13-10
平行四边形机构 13-4
评定长度 13-10
屏蔽 19-6
屏蔽双绞线 14-6
屏幕分辨率 8-1
剖分式滑动轴承 13-9
普通V带 13-7
普通二极管 4-1
普通平带 13-7
普通平键 13-8
普通型分散控制器 14-3
瀑布式模型 19-5

Q

七段数码管显示器 2-7
七段显示器 21-4
齐次方程 20-9
齐格勒-尼可尔斯法则 20-7

奇异函数 5-7
奇异矩阵 20-9
企业建筑物集成 14-6
启动 21-6
启动计数 2-5
起点 20-4
起动电流 3-3
起动过程 18-2
起动转矩 3-4
起动转矩 3-7
起始地址 2-3
起始角度 17-2
起始位 2-6
气动阀（横隔膜）10-2
气动量仪 1-4
气动执行器 12-1
气鼓 10-4
气关阀 10-2
气开阀 10-2
气敏元件 1-4
气提塔 10-4
气体激光器 1-5
气隙 3-3
器件封装 17-2
器件换流 4-5
千赫 5-1
千瓦时 5-9
牵出同步 3-9
牵出转矩 3-9
牵入同步 3-9
牵入转矩 3-9
前馈控制 10-4
前馈控制 20-3
前馈校正 20-6
前馈作用 10-4
前连接器模块 6-3
前连接器模块 6-3
前向差错控制 9-4
前向纠错方式 12-5
前置放大 19-2
箝位 4-1
箝位二极管 4-1
欠补偿 18-1
欠励磁 3-9
欠阻尼 20-3
欠阻尼的 10-3

嵌入式符号 8-1
嵌入式软件 19-5
嵌入式实时操作系统 19-5
嵌入式系统设计 17-1
嵌入性 13-9
强度调制 1-5
强迫关断电路 18-1
强迫换流 4-5
强迫通风冷却 3-5
强迫响应 5-7
强制分量 5-7
强制风冷 4-1
强制通风 10-4
强制值 6-4
桥式可逆斩波电路 4-3
桥式整流电路 22-10
切除PCB板边角 17-5
切向键 13-8
切向力 3-2
切削 3-2
倾斜度 13-10
清除发送 12-5
清零 21-5
擎住电流 4-1
擎住效应 4-1
请求 15-3
请求发送信号 12-5
请求帧 9-6
求和电路 22-2
球笼式同步万向联轴器 13-8
球面副 13-3
球面凸轮 13-5
球轴承 13-9
球轴承 3-7
区域内对象的选择 17-3
区域外对象的选择 17-3
曲柄 13-4
曲柄摆动导杆机构 13-4
曲柄存在条件 13-4
曲柄导杆机构 13-4
曲柄滑块机构 13-4
曲柄摇杆机构 13-4
曲柄移动导杆机构 13-4
曲线回归方程 16-6

曲线拟合 19-4
驱动 2-2
驱动点导纳 5-14
驱动点阻抗 5-14
驱动电路 4-1
驱动方程 21-5
驱动器 12-4
屈服极限 13-2
屈服强度 13-2
趋肤效应 3-9
趋势 8-1
取样-保持电路 21-10
取样长度 13-10
取样性质 5-7
取址 2-1
去磁效应 3-1
去极值 19-4
全波整流电路 4-2
全补偿 overall 18-1
全齿高 13-6
全封闭 3-7
全回流 10-4
全加器 21-4
全控型器件 4-1
全屏显示 17-1
全桥电路 4-2
全桥整流电路 18-1
全双工 2-6
全双工方式 12-5
全跳动 13-10
全响应 5-7
全压起动 18-5
权 16-2
权电流 21-10
权电阻 21-10
确认 15-3
确认测试 19-5
群控电梯 14-3

R

燃料供给 10-4
燃料喷嘴 10-4
燃料输送泵 10-4
燃料油 10-4
绕线转子 3-7
绕线转子异步电动机 3-7
绕组 3-3

热备份 8-1
热备份管理器 8-1
热插拔 19-3
热磁效应 1-3
热电极 1-3
热电偶 10-4
热电桥 1-3
热电效应 1-3
热电阻 1-3
热击穿 4-1
热继电器 3-8
热键 17-5
热交换器 10-4
热敏电阻器 1-3
热敏开关 1-3
热敏元件传感器 1-3
热能 3-2
人机对话 19-3
人机接口 8-1
人机界面 11-1
人字齿轮 13-6
任意跳转 6-4
日光灯 3-3
容错技术 12-6
容抗 5-9
容纳 5-9
容器 10-4
容性 3-5
容性阻抗 5-9
容许极限 4-2
溶剂 10-4
熔断器 3-8
冗余 8-1
冗余的处理器 11-2
冗余设计 19-6
柔性冲击 13-5
入口地址 2-4
入射角 20-4
软件 2-1
软件测试 19-5
软件调试 12-10
软件开发过程 12-10
软件可靠性设计 19-6
软件设计 19-5
软件陷阱 19-6
软件延时法 12-4

软件延时方式 12-2
软开关 4-7
弱磁控制 18-2

S

塞尔维斯特 20-9
三表法 5-12
三次谐波磁通 3-9
三端元件 5-2
三极管 21-3
三极管-三极管电路 21-3
三角波 18-6
三角波变锯齿波 22-9
三角波发生电路 22-9
三角形电阻网络 5-2
三角形分布 16-2
三角形接法 3-5
三阶最佳 18-2
三态门 21-3
三铁心柱 3-5
三通阀 10-4
三维建模 7-1
三相 3-5
三相半波可控整流电路 4-2
三相电路 5-12
三相电源 5-12
三相对称电路 3-5
三相母线 7-3
三相桥式可控整流电路 4-2
三相三线制 5-12
三相四线制 3-8
三相制 5-12
散热 22-8
扫描填充 17-2
扫描周期 6-2
筛分性质 5-7
删除线号 7-2
闪蒸罐 10-4
扇出 21-3
扇出式布线控制 17-6
扇入 21-3
扇形图 17-2
商业楼宇自动化 15-2
上拉电阻 2-5
上偏差 13-10

上升时间 20-3
上升沿 21-5
上限截止频率 22-4
上限阈值电压 21-7
上游过程 10-1
蛇形弹簧联轴器 13-8
设备 19-3
设备发现 15-4
设备间 14-6
设备类型 15-3
设备描述 14-2
设备描述符 15-4
设备描述语言 14-2
设备名称 6-1
设备驱动 9-3
设备声明 15-4
设定值 10-3
设定值控制 10-1
设计规则和约束编辑器 17-6
设计规则检查 17-2
设计规则检验 17-5
设计规则约束 17-5
设计数据库 17-2
设计原点 17-2
设置选项 15-6
设置元件时切断导线 17-2
设置粘贴阵列 17-7
射级跟随 22-4
射级偏置电路 22-4
射极跟随器 22-4
射极耦合 21-3
射频识别技术 15-2
摄动 20-7
深度负反馈的特点 22-7
深度负反馈电路 22-7
深沟球轴承 13-9
神经网络 12-9
神经网络和模糊控制 11-3
神经元 12-5
神经元专用芯片 12-5
升程 13-5
升降压变压器 4-3
升压变压器 3-5

升压电路 4-3
升压斩波电路 4-3
生产过程 12-1
生产计划与调度 12-1
生存周期 19-5
省煤器 10-4
剩磁 3-1
失步 3-9
失调电流 22-6
失调电压 22-6
失效率 19-6
施工图 10-4
施密特触发器 21-7
湿度传感器 14-2
十进制 2-1
十六进制 2-1
十字滑块机构 13-4
十字滑块联轴器 13-8
石英传感器 1-2
石英晶体 22-9
石英晶体振荡器 21-7
时变系统 10-3
时不变系统 12-2
时间常数 5-7
时间继电器 6-1
时间模型 19-6
时间延迟 10-3
时间邮票 19-5
时序 2-2
时序电路 21-5
时序仿真 17-2
时序图 21-1
时域 10-3
时域函数 5-13
时域响应 20-3
时钟触发器 21-5
时钟存储器 6-4
时钟信号 21-5
时钟周期 2-2
实部 3-5
实根 20-3
实际变压器 3-5
实际尺寸 13-10
实际啮合线长度 13-6
实际偏差 13-10
实际要素 13-10

索 引

实际直流电流源 5-9
实际直流电压源 5-9
实际装置 10-4
实时分布式控制 11-1
实时交互参考 7-3
实时趋势 8-1
实时软件 19-5
实时时钟 12-1
实时数据库 14-2
实数 2-1
实数单根 5-13
实体设计 17-2
实心销轴 13-8
实验对比法 16-2
实值 3-5
实轴 20-4
史密特触发器 2-2
使能端 21-4
使能输入/输出 6-4
使用或取消电气栅格 17-2
事故电流 4-1
事故过电压 4-1
事件 8-1
事件描述 8-1
事件优先权 8-1
事务处理软件 19-5
势垒电容 22-3
势能 3-2
视频点播 14-6
视图自动移动功能 17-2
视在功率 3-5
试验装置 10-4
适配器电缆 6-2
收敛 20-10
收敛因子 5-13
手册、指南 21-6
手持编程器 6-2
手动控制 12-7
手动状态 8-1
寿命 3-4
寿命系数 13-9
受控变量 10-3
授权 8-1
授权管理 17-1
授权激活 17-1

授权协议 17-1
输出 5-1
输出电路 6-3
输出电压 5-5
输出电阻 22-4
输出方程 20-9
输出幅度 12-8
输出功率 3-2
输出缓冲器 2-2
输出量 20-1
输出流量 10-2
输出逻辑宏单元 21-9
输出设备 2-7
输出输入快 21-9
输出特性 22-3
输出文件路径 17-2
输出限幅 18-2
输出允许 2-2
输入 5-1
输入/输出参数 6-4
输入电压 5-5
输入电阻 5-2
输入功率 3-2
输入缓冲器 2-2
输入流量 10-2
输入设备 2-7
输入输出端口 2-2
输入输出接口 6-2
输入-输出描述 20-2
输入输出模块 11-2
输入输出设备 2-7
输入阻抗 5-5
属性 15-3
鼠标轮设置 17-2
鼠标 2-1
鼠笼 3-7
鼠笼转子异步电动机 3-7
树 5-3
树支 5-3
数/模 14-3
数据包 9-5
数据编码 9-2
数据编码和传输 9-7
数据变化脚本程序 8-1
数据采集 19-2
数据采集单元 11-2

数据采集器 14-3
数据采集系统 12-8
数据采集与监控系统 8-1
数据操作指令 6-4
数据处理 19-4
数据传送方式 12-5
数据存储器扩展 2-2
数据发送位 2-6
数据缓冲器 2-1
数据结构 19-5
数据库系统 14-3
数据块 6-4
数据块寄存器 6-4
数据类型 8-1
数据类型声明 6-4
数据链路层 9-1
数据模型 19-6
数据请求 15-3
数据区 2-2
数据锁存信号 2-2
数据通信 9-1
数据通信设备 12-5
数据通信适配器 14-6
数据位 2-6
数据位低位 2-6
数据位高位 2-6
数据信号 2-2
数据载波检测信号 12-5
数据帧 2-6
数据帧的路由 15-5
数据终端就绪信号 12-5
数据终端设备 9-1
数据准备就绪 12-5
数据字节 DBB 6-4
数据总线 2-2
数码比较器 21-4
数码管 12-3
数码显示器 21-4
数码序列 20-8
数模转换 2-2
数模转换器 2-8
数学模型 20-2
数字的 17-2
数字传感器 19-2
数字地 12-2
数字控制器 12-7

数字控制算法 12-7
数字量 2-8
数字量报警 8-1
数字量输出 14-3
数字量输出通道 12-2
数字量输入 14-3
数字量输入/输出模块 6-3
数字量输入通道 12-2
数字滤波 12-6
数字式电液控制系统 1-4
数字式转速传感器 12-4
数字信号 2-2
数字用户单元 14-6
数组元素 6-4
刷握 3-3
衰减 20-3
衰减比 10-3
衰减常数 5-7
衰减特性 20-6
衰减系数 18-2
双闭环调速系统 18-2
双刀双掷开关 3-8
双端电路 4-8
双工 SC 连接器 14-6
双滑块机构 13-4
双积分 21-10
双积分型模数转换器 19-2
双极结型晶体管 4-1
双绞线电缆 11-4
双精度整数 6-4
双列直插封装 17-7
双列直插式封装 21-3
双列轴承 13-9
双面印制板 17-5
双曲柄机构 13-4
双输入单输出 12-9
双速电机 3-8
双头蜗杆 13-6
双稳态 18-2
双向的 2-2
双向二极晶闸管 4-1
双向晶闸管 4-1
双向可控硅 12-4
双向数据线 2-2

双向推力轴承 13-9
双向移位寄存器 21-6
双摇杆机构 13-4
双重零点 20-7
双转动导杆机构 13-4
双字 2-1
双作用离合器 13-8
水轮发电机 3-9
水平尺寸 8-1
水平导线 7-2
水平等间距排列 17-3
顺接/正向串联 5-10
顺流 10-4
顺时针 3-3
顺时针方向 5-9
顺时针旋转 8-1
顺时针旋转被选对象 17-3
顺序查表法 12-6
顺序功能图 6-5
瞬时电压 18-1
瞬时功率 5-9
瞬态响应 20-2
说明 17-4
丝印层 17-5
死点 13-4
死区 20-10
四代技术 19-5
四端元件 5-10
四象限运行 3-4
伺服电动机 3-10
松边拉力 13-7
松耦合 3-5
送风机 14-3
速度调节器 20-1
速度反馈 18-1
速度环 18-2
速度控制 12-7
塑性变形 13-2
算法 19-4
算术逻辑单元 21-4
算术平均 19-4
算术平均滤波 12-6
算术校验和 9-4
算术运算 2-3
算数平均值 16-2

算数运算指令 6-4
随动系统 12-8
随机采样 12-2
随机存储器 21-8
随机存取存储器 2-1
随机干扰 19-4
随机扰动 20-1
随机误差 16-2
随机误差的合成 16-3
随机误差特征 16-2
缩放精度 17-1
缩小 8-1
锁存 2-2
锁存器 2-2

T

他励 3-3
塔板 10-4
钛合金 13-2
泰勒级数 20-10
泰勒级数展开 10-3
炭刷 3-3
弹出窗口 8-1
弹簧管 1-7
弹性挡圈 13-8
弹性极限 13-2
弹性联轴器 13-8
弹性元件 1-7
弹性圆柱销 13-8
弹性柱销齿式联轴器 13-8
碳钢 13-2
陶瓷温度传感器 1-4
套筒联轴器 13-8
特定规范 15-3
特定谐波消去法 4-6
特解 5-7
特勒根定理 5-4
特勒根功率定理 5-4
特勒根似功率定理 5-4
特殊功能寄存器 2-2
特性表 21-5
特性方程 21-2
特性分析 1-2
特性曲线 22-4
特征多项式 20-2
特征方程 10-3

特征根 5-7
特征频率 22-4
特征向量 20-9
特征值 20-9
特种电机 3-10
梯形图 6-2
梯形图编辑器 6-4
提取 22-1
体积流量 1-7
体系结构 15-1
替代窗口 8-1
替代定理 5-4
替换元件 7-3
添加电源平面 17-5
添加横档 7-2
添加信号层 17-5
填充域 17-2
条件编译 15-6
条件分支 8-1
条件积分 12-7
调幅 21-1
调节阀 10-2
调节阀流量特性 10-2
调节阀流通能力 10-2
调节阀增益 10-2
调节过程 12-7
调节器整定 10-3
调节时间 18-2
调频 1-2
调试 8-1
调试及实验 12-10
调试器 15-6
调速电机 18-5
调速范围 18-1
调速系统 18-1
调速指标 18-1
调心滚子轴承 13-9
调心轴承 13-9
调压调速系统 18-1
调整曲线 3-3
调制波 18-6
调制度 4-6
调制技术 19-2
调制解调 19-3
调制解调器 2-6
调制器 20-8

跳动公差 13-10
跳转目标 6-4
铁磁材料 3-1
铁损耗 3-4
铁芯 3-3
铁芯线圈 5-10
停止计数 2-5
停止位 2-6
停止系统模式 6-2
通带电压放大倍数 22-9
通带截止频率 22-9
通道与接口 12-2
通孔 17-5
通频带 22-4
通态 18-4
通态(导通状态) 4-6
通态电流临界上升率 4-1
通态电压 4-1
通态电压临界上升率 4-1
通态损耗 4-3
通信插座 14-6
通信技术 19-1
通信链路 9-1
通信网络 14-6
通信自动化系统 14-1
通讯 2-6
通讯方式 2-6
通讯功能块 6-6
通用串行总线 19-3
通用串行总线 9-3
通用电动机 18-1
通用逻辑快 21-9
通用型 22-6
通用异步收发器 9-2
通用阵列逻辑器件 21-9
同步变压器 18-1
同步触发器 21-5
同步带传动 13-7
同步电动机 3-9
同步电感 3-9
同步电抗 3-9
同步调制 4-6
同步时序电路 21-6
同步速 3-6
同步通信 12-5
同步通讯 2-6

索 引

同步整流电路 4-2
同或 21-2
同名端 3-5
同相 3-5
同相输入 22-2
同相输入端 22-6
同相输入端 5-5
同轴度 13-10
铜电阻 1-3
铜合金 13-2
铜损耗 3-4
透明度 17-1
透明浮动窗口 17-1
凸极 3-7
凸极效应 3-9
凸轮廓形 13-5
凸轮理论廓线 13-5
凸轮轮廓 13-5
凸轮轮廓上的尖点 13-5
凸缘联轴器 13-8
图 5-3
图标菜单 7-4
图解法 20-10
图解分析法 22-4
图腾柱 21-3
图形编辑 17-3
图形符号 21-5
图形工具箱 8-1
图形接口 8-1
图形显示 17-2
图纸连接端口的信号流向 17-2
推荐宽度 17-6
推力轴承 13-9
推挽 22-8
推挽变流器 4-8
退火 13-2
拖动 3-1
拖动转矩 3-8
拖拽被选的对象 17-3
拓扑 9-3

W

瓦 3-1
瓦特 5-1
瓦特表 5-9
外部存储器 2-2

外部地址总线 2-2
外部控制总线 2-2
外部链接 8-1
外部输入脉冲 2-5
外部数据总线 2-2
外部通用设备 12-1
外部中断 2-4
外径 3-7
外圈 13-9
外设输出双字 6-3
外特性 22-2
外围的 2-2
外形尺寸（滚动轴承）13-9
完全配方法 5-13
完整功能设备 15-3
万向联轴器 13-8
万用表 4-1
网关 9-1
网际互连 9-7
网孔 5-3
网孔电流 5-3
网孔电流法 5-3
网孔分析法 5-3
网络 12-5
网络标签 17-1
网络表 17-3
网络表选项 17-2
网络操作系统 9-7
网络层 9-5
网络层管理实体 15-5
网络层数据实体 15-5
网络地址 15-3
网络端口 17-1
网络发现 15-5
网络函数 5-11
网络化传感器 19-1
网络接口单元 14-3
网络控制系统 14-3
网络通信指令 6-4
网络信息库 15-5
网络形成 15-5
网络终端设备 14-3
网络组态 6-6
网络最大深度 15-5
网状网络 15-1

威尔逊电流源 22-6
微处理器 2-1
微法 5-6
微分电路 22-2
微分方程 10-3
微分负反馈 18-2
微分环节 20-2
微分控制 18-1
微分时间 12-7
微分型 21-7
微分性质 5-13
微观不平度十点高度 13-10
微亨 5-6
微机电系统 15-1
微商 20-8
微小误差的取舍准则 16-3
微型计算机 2-1
韦伯 5-10
维持电流 4-1
维护 19-5
维数 20-9
尾码 2-1
未定系统误差的合成 16-3
未定义 21-5
位 2-1
位、字节和字符 9-2
位存储器地址 2-3
位存储器地址区 6-4
位累加器 6-4
位寻址 2-3
位移 3-2
位移测量 1-2
位移传感器 1-6
位移定理 20-8
位指令 6-4
位置变送器 1-2
位置代号 7-4
位置度 13-10
位置公差 13-10
位置式 12-7
位置随动系统 20-1
温标 1-7
温标 3-2

温差电流 1-3
温度变送器 1-7
温度补偿 1-2
温度补偿电路 1-4
温度传感器 1-3
温度检测 1-7
温度误差 1-2
温度系数 1-3
温漂 1-1
温升 3-3
文本编辑器 17-1
文档 8-1
文件夹路径 8-1
文件类型 17-1
文件输出选项 17-2
文件锁定 17-1
稳定 18-2
稳定判据 22-4
稳定条件 18-2
稳定性 10-3
稳定裕度 22-4
稳定裕量 20-1
稳态 10-1
稳态跟随误差 18-1
稳态结构图 18-1
稳态精度 20-1
稳态抗扰误差 18-1
稳态速降 18-1
稳态误差 20-1
稳态响应 5-7
稳态值 20-1
稳压电路 22-3
稳压电源 22-10
稳压二极管 22-3
稳压管稳压电路 22-10
涡街流量计 1-7
涡流 3-1
涡流传感器 1-4
蜗杆 13-6
蜗杆与蜗轮 13-6
蜗轮 13-6
蜗轮副 13-6
蜗轮滚刀 13-6
握手信号 2-2
无齿侧间隙啮合方程式 13-6

无符号数 2-3
无符号数加法 2-3
无符号数减法 2-3
无功分量 5-9
无功伏安/乏 5-9
无功功率 3-5
无功功率表 3-5
无关项 21-2
无级变速 13-7
无记忆元件 5-6
无量纲的 10-2
无扰动切换 12-7
无刷 3-7
无刷励磁系统 3-9
无稳态多谐振荡器 1-3
无线传感器 15-1
无线个域网 15-1
无线数据传输 19-3
无线自组网按需平面距离矢量路由协议 15-2
无效状态 21-6
无源/零输入 RC 电路 5-7
无源/零输入 RL 电路 5-7
无源电路元件 5-6
无源滤波器 5-11
无源逆变 4-5
无源线性电路 5-11
无源线性元件 5-6
无源元件 5-1
无阻尼 20-3
无阻尼自然频率 20-3
物理层 9-1
物理量检测仪表 1-1
物料平衡 10-2
物位检测 1-7
误差 16-1
误差方程 16-5
误差分配 16-3
误差信号 20-3
误差因数 1-2
误码率 9-2

X

西门子 3-1
吸入流量 10-4

吸收电路 4-1
吸收电容 4-1
吸收功率 3-5
系统辨识 12-1
系统框图 20-2
系统平台 8-1
系统软件 19-5
系统设计 12-10
系统稳定性 18-1
系统误差 16-2
系统误差的合成 16-3
系统误差特征 16-2
系统字体 17-2
下降时间 21-1
下降沿 21-5
下拉菜单 7-1
下拉电阻 2-2
下拉列表 7-4
下偏差 13-10
下限截止频率 22-4
下限阈值电压 21-7
下游过程 10-1
先由上至下,再按由左至右 17-3
先由下至上,再按由左至右 17-3
先由左至右,再按由上至下 17-3
先由左至右,再按由下至上 17-3
显示 2-2
显示程序 2-7
显示缓冲区 2-7
显示或隐藏栅格 17-2
显示接口 2-7
显示器 2-1
显示器设计技术 2-7
显示文件结构 17-1
显示项目结构 17-1
显示仪表 12-3
显示栅格 17-1
显示子程序 2-7
显著性检验 16-6
现场可编程逻辑阵列 21-9
现场可编程门阵列 21-9

现场控制单元 11-2
现场控制总线 10-4
现场总线隔离变压器 6-3
现场总线基金会 12-5
现场总线控制系统 12-1
现代控制理论 20-1
现代仪器 19-1
现态 21-6
线参考 7-1
线电流 3-5
线电压 3-5
线反转 19-3
线号 7-2
线轮廓度 13-10
线圈 6-4
线圈节距 3-6
线圈匝数 5-10
线网络 17-2
线形固态图像传感器 1-4
线性变化的系统误差 16-2
线性变换 11-3
线性标度变换 8-1
线性标注 17-5
线性参数 16-5
线性递推回归 16-6
线性电感元件 5-6
线性电路 5-1
线性电容元件 5-6
线性电阻 5-1
线性电阻元件 5-1
线性度 10-2
线性度误差 19-2
线性对称电路 5-4
线性阀 10-2
线性化 1-1
线性控制系统 20-1
线性模型 10-3
线性区 5-5
线性时不变 20-7
线性时不变系统 20-9
线性网络 5-14
线性系统稳定判据 10-3
线性小信号等效电路 22-3

线性性质 5-13
线性元件 5-6
线性阵列 17-7
线性值 5-4
线性组合 5-4
线与 21-3
限幅 22-9
限幅电路 18-2
限幅滤波 12-6
限幅器 18-1
限流保护 18-1
限流装置 18-1
限速滤波 12-6
限位开关 6-1
相电流 3-5
相电压 3-5
相对精度 19-2
相对湿度 1-4
相对稳定性 20-5
相对误差 16-1
相对于 3-5
相对增益 10-4
相关分析 19-4
相关关系 16-6
相关系数 16-3
相关指数 16-6
相加点 20-2
相角 3-5
相角超前量 20-6
相角条件 20-6
相角滞后量 20-6
相控 4-2
相量 3-6
相量法 5-8
相量图 3-5
相邻项 21-2
相频特性 22-4
相频图 20-5
相平面法 20-10
相位 20-5
相位差 5-8
相位调制 1-5
相位交界频率 20-5
相位控制 18-1
相位裕度 22-4
相位裕量 20-5

相序 3-5
相移 3-5
相移键控法 12-5
香农定理 20-8
响应 15-3
响应曲线 12-8
响应曲线法 12-7
响应时间 6-6
响应帧 9-6
向导 8-1
向高位进位 21-4
向心滚子轴承 13-9
向心力 3-4
向心球轴承 13-9
项目 8-1
项目管理器 7-1
项目面板 17-1
项目设置 7-1
项目特性 7-1
像素 8-1
橡胶 13-2
橡胶板联轴器 13-8
橡皮图章工具 17-3
消除干扰 19-6
消防给水 14-4
消防设备 14-4
消防栓 14-4
消防用水 14-4
消防自动化 14-1
消防自动化系统 14-4
销 13-8
销轴 13-8
小功率直流电动机 12-4
小规模集成电路 21-4
小数/分数 2-1
小信号模型 22-3
小型计算机 2-1
肖特基二极管 4-1
肖特基势垒二极管 4-1
效率 3-2
楔 13-8
协处理器 2-2
协调器 15-3
协议 12-5
协议分析仪 15-6
协议栈 15-1

协议栈规范 15-3
斜率 10-2
斜坡（信号）10-3
斜坡响应 20-2
斜坡信号 20-3
谐波 3-6
谐波齿轮传动 13-6
谐波电压 3-6
谐波分量 20-10
谐波分析 18-1
谐波响应 20-2
谐波效应 3-9
谐振 4-7
谐振电路 5-11
谐振电路阻抗 5-11
谐振峰值 20-5
谐振频率 5-11
谐振曲线 1-2
谐振直流环电路 4-7
谐振阻抗 5-11
写允许 2-2
新文档默认项 17-1
信标 15-3
信道 15-3
信道监听 15-2
信号地 17-2
信号调理 19-2
信号调制 12-5
信号箭头 7-2
信号流图 20-2
信号模块 6-3
信号衰减 9-4
信号完整性 17-6
信号源 22-1
信号转换电路 22-9
信息插座 14-6
信息格式 9-6
信息技术 19-1
信息库维护 15-5
信噪比 22-3
信噪比 9-2
星形电阻网络 5-2
星型、环型和总线型拓扑 9-7
星型接法 3-5
行波 3-6

行程速度变化系数 13-4
行传输 1-4
行列式 20-9
行扫描 19-3
行向量 20-9
行星齿轮 13-6
行星轮系 13-6
行星转臂 13-6
行选择线 21-8
行注释 6-4
形状公差 13-10
性能扩展 22-6
性能指标 20-1
修改过的文档 17-2
修改阶梯 7-2
修剪导线 7-2
修正值 16-1
虚部 3-5
虚地 22-2
虚短 22-2
虚拟背板总线 6-2
虚拟链路 9-5
虚拟现实 14-1
虚线 7-3
虚约束 13-3
虚轴 20-4
续流二极管 18-1
续流元件 4-1
蓄电池 4-3
旋转被选的对象 17-3
旋转变压器 18-1
旋转磁场 3-6
旋转磁通 3-6
旋转方向 3-6
选频 22-9
选频特性 5-11
选频网络 22-9
选通 2-2
选通信号 2-2
选项 8-1
选择PCB板轮廓 17-5
选择电路板的层数 17-5
选择接口控制 12-5
选择开关 7-3
选择模板 7-1
学生分布 16-2

学习控制系统 12-1
寻址错误 6-3
寻址方式 2-3
循环程序处理 6-4
循环初态 2-3
循环结束 2-3
循环结束条件 2-3
循环控制 6-4
循环码 1-6
循环冗余校验 9-4
循环体 2-3

Y

压电传感器 1-2
压电式加速度计 1-2
压电式流量计 1-2
压电式压力传感器 1-2
压电陶瓷 1-2
压电效应 1-2
压控电压源 22-9
压控振荡器 18-6
压力传感器 1-2
压力检测 1-7
压力降 10-2
压敏电桥 1-2
压敏电阻 1-2
压敏二极管 1-2
压缩机 10-4
压阻式传感器 1-2
压阻系数 1-2
压阻效应 1-2
牙嵌式离合器 13-8
牙嵌式联轴器 13-8
雅可比矩阵 20-9
烟道 10-4
烟气 10-4
烟雾报警器 14-4
延迟 21-5
延迟失真 9-4
延迟时间 20-3
延迟性质 5-13
延时闭合 3-8
延时断开 3-8
延时-功耗积 21-3
沿检测 6-4
演化 19-5
验收测试 19-5

扬程 10-4
阳极 2-2
摇杆 13-4
摇块机构 13-4
遥控 3-10
要显示的对象 17-1
页面标志 17-1
页面端口 17-1
页面接口 17-1
液晶 2-7
液晶显示器 12-3
液体动压滑动轴承 13-9
液体静压滑动轴承 13-9
液位 10-2
液位变送器 1-4
液位传感器 14-2
一端口网络 5-14
一阶保持器 12-8
一阶常微分方程式 5-7
一阶电路 5-7
一阶反应 10-4
一阶惯性环节 10-3
一阶过程 10-3
一阶极点 5-13
一阶系统 20-3
一阶因子 20-5
一元非线性回归 16-6
一元线性回归 16-6
一元线性回归方程 16-6
一兆(百万) 2-2
仪表 10-2
仪表装置 12-5
仪器 19-1
仪器和控制 6-1
仪器精密轴承 13-9
仪用放大器 19-2
移动被选的对象 17-3
移动从动件 13-5
移动端子 7-4
移动副 13-3
移动回路 7-3
移动凸轮 13-5
移位和循环指令 6-4
移位寄存器 21-6
移相全桥电路 4-2
已安装的库 17-1

已定系统误差的合成 16-3
以阵列形式粘贴 17-3
异步传输模式 14-1
异步电动机 3-6
异步电机 3-6
异步调制 4-6
异步和同步系统 9-2
异步时序电路 21-6
异步通信 12-5
异步通讯 2-6
异或 21-2
异相 3-5
抑制电磁干扰 19-6
译码法编址 2-2
译码器 2-2
溢出 2-2
溢出标志 2-2
溢出中断 2-4
因变量 10-3
阴极 4-1
阴燃 14-4
音箱 2-1
引出点 20-2
引脚 17-1
引脚的信号流向 17-2
引脚属性 17-4
引入负反馈的一般原则 22-7
引用误差 16-1
隐藏栅格 17-1
隐极 3-7
隐极转 3-9
印刷电路板编辑器 17-1
印制板电路 17-1
应变极限 1-2
应变片 1-2
应变式压力传感器 1-2
应答 2-2
应答信号 2-7
应力传感器 1-2
应力极限 13-2
应用层 9-1
应用程序窗口 8-1
应用程序存储器 6-2
应用程序框架 15-1

应用软件 19-5
应用系统标识符 8-1
应用支持子层 15-3
英文字母 17-2
英制单位 17-2
荧光辐射式温度传感器 1-3
荧光数码管 21-4
硬件 2-1
硬件调试 12-10
硬件接线 6-2
硬件可靠性设计 19-6
硬件译码 12-3
硬件组态 6-3
硬开关 4-7
硬链接 21-9
硬盘 2-1
永磁铁 3-1
用户连接器 14-6
用户描述符 15-4
用户信息 17-1
用一条语句进行的 DB 访问 6-4
优化控制 10-4
优化连线 17-2
优先编码器 21-4
优先级 2-4
优先级等级 6-4
优先配置 13-10
优先数系 13-10
油槽 13-9
油浸 3-5
油孔 13-9
油箱 3-5
有伴电流源 5-3
有伴电压源 5-3
有符号数 2-3
有功分量 18-5
有功功率 3-5
有静差调速系统 18-1
有理数 20-5
有理真分式 5-13
有效负载 18-5
有效拉力 13-7
有效数字 16-1
有效值 3-5

有源负载 22-6
有源功率滤波器 4-5
有源滤波器 1-2
有源逆变 4-5
有源器件 4-5
有源下拉电路 21-3
有源元件 5-1
右半平面 20-3
右对齐 8-1
右手定则 3-6
右手螺旋定则 5-6
余 3 码 21-1
余弦加速度运动规律 13-5
与 21-2
与安全相关 11-4
与非 21-2
与或非 21-2
语句表 6-4
语言变量 12-9
语言系统标识符 8-1
阈值 10-2
阈值电平 21-7
阈值电压 4-5
元件 17-7
元件安置 17-2
元件标号 17-2
元件标记 7-3
元件标记格式 7-1
元件面 17-5
元件描述 7-3
元件引脚编辑器 17-4
元件属性 17-2
元件注释 7-3
原边 3-5
原边漏电抗 3-5
原点 17-2
原电路 5-4
原动机 3-9
原函数 5-13
原理 5-4
原理图 3-3
原理图标准 17-2
原理图参数设置 17-2
原理图符号 7-3
原理图库 17-4

索引

原料 10-4
原码 2-1
原形开发模型 19-5
原语 15-3
圆带传动 13-7
圆顶罐 10-4
圆度 13-10
圆焊盘 17-5
圆跳动 13-10
圆柱度 13-10
圆柱副 13-3
圆柱滚子轴承 13-9
圆柱凸轮 13-5
圆柱蜗杆 13-6
圆柱销 13-8
圆锥滚子轴承 13-9
圆锥销 13-8
圆锥形凸轮 13-5
源 2-1
源绑定 15-4
源操作数 2-3
源地址域 15-5
源极 22-5
源箭头 7-2
源路由子域 15-5
源信号 7-2
远程 I/O 子系统 6-2
远程引用 8-1
远程终端 9-6
远程桌面协议 8-1
约当标准形 20-9
约束 13-3
约束值 10-4
越限报警 12-4
云母 3-3
允许接收位 2-6
允许设备连接 15-5
运动副 13-3
运动副元素 13-3
运动确定性 13-3
运动失真 13-5
运放 22-6
运算导纳 5-13
运算电路 5-13
运算法 5-13
运算放大器 5-5

运算阻抗 5-13
运行 15-6
运行控制位 2-5
运行系统 8-1

Z

匝数 3-1
匝数比 5-10
杂散损耗 3-5
载波 18-4
载波比 4-6
载波监听多路访问/碰撞检测 9-7
载波频率 4-6
载波信号 20-8
载波侦听多路访问/介质访问控制 12-5
载波侦听多路访问冲突避免 15-2
载流子 22-3
再生(刷新) 21-8
在系统可编程 21-9
在线 12-10
暂态/瞬变状态 5-7
暂态分量 5-7
暂态分析 5-7
暂态响应 5-7
噪声容限 21-3
噪声信号 20-7
噪声源 19-6
增补 8-1
增量式 12-7
增强型 22-5
增速比 13-7
增益 20-2
增益交界频率 20-5
增益误差 19-2
增益裕量 20-5
闸阀 10-4
闸流管 4-1
栅格 17-2
栅极 22-5
栅极偏置电路 22-5
窄 V 带 13-7
斩波控制 18-4
展开系数 5-13
展伸不确定度 16-4

占空比 12-4
占位模块 6-3
张紧轮 13-7
照明控制器 14-3
照明系统 14-3
罩极 3-10
遮光路式温度传感器 1-3
折合 3-5
折线 17-2
折线化模型 22-3
真值 16-1
真值表 21-2
振荡 18-2
振荡环节 20-2
振荡频率 2-5
振荡特性 10-3
振荡条件 22-9
振荡周期 21-7
振幅 10-2
振幅测量 1-2
振簧式频率传感器 1-6
振铃现象 12-8
振铃指示信号 12-5
蒸发器 10-4
蒸汽锅炉 10-4
整距 3-3
整流 4-2
整流变压器 18-1
整流电路 18-1
整流二极管 4-2
整流滤波器 18-1
整流平均值 18-1
整流器 4-2
整流纹波系数 18-1
整数 2-1
整体设计过程 22-1
整体式滑动轴承 13-9
正比于 3-5
正边沿 21-5
正常齿高制 13-6
正常全齿高 13-6
正电荷 5-1
正电源 5-5
正定 20-9
正反馈 22-7
正规方程 16-5

正号 21-1
正火 13-2
正激电路 4-8
正交拖动 17-2
正逻辑 21-3
正脉冲 2-5
正态分布 16-2
正弦 PWM 4-6
正弦波 3-6
正弦波振荡电路 22-9
正弦电流 3-5
正弦函数 5-8
正弦机构 13-4
正弦加速度运动规律 13-5
正弦量 5-9
正弦输入 10-3
正弦稳态 5-9
正弦稳态响应 5-9
正弦信号 20-3
正相序 3-5
正向电流 4-1
正向偏置 4-1
正向偏置安全工作区 4-1
正向平均电流 4-1
正向压降 4-1
正向转动 12-4
正组触发装置 18-3
正作用 12-7
帧 2-6
帧传输 1-4
帧格式 2-6
帧控制域 15-5
帧类型子域 15-5
帧校验 15-2
支路 5-3
支路电流 5-3
支路电流法 5-3
支路电压 5-3
知识产权 21-9
执行 2-1
执行顺序 17-3
执行元件 10-3
直齿圆锥齿轮 13-6
直交直电路 4-8
直角坐标形式 5-8

直接 I/O 访问 6-3
直接程序设计法 12-8
直接存储器存取 12-2
直接电流控制 4-8
直接耦合放大电路 22-4
直接起动电动机 18-5
直接数字控制 11-2
直接数字控制器 14-2
直接序列扩频 15-2
直接寻址 2-3
直接直流变换电路 4-8
直径系列(滚动轴承) 13-9
直流 18-1
直流电机 18-1
直流电流 3-3
直流电流源 5-1
直流电气传动 18-1
直流电压源 5-1
直流电源 5-2
直流电阻 5-9
直流调速系统 20-2
直流互感器 18-1
直流继电器 12-4
直流励磁发电机 DC 18-1
直流模型 22-4
直流他励电动机 20-2
直流他励发电机 20-2
直流斩波 4-3
直流斩波电路 4-3
直流斩波器 18-1
直流-直流变换器 4-3
直通式光纤连接器 14-6
直线 3-4
直线度 13-10
直轴 3-9
止推滑动轴承 13-9
只读存储器 2-1
指令 2-3
指令表 6-2
指令周期 2-2
指示 15-2
指数 10-3
指数电路 22-2
指数规律衰减 5-7

指数函数 20-9
指数形式 5-8
制动 3-2
制动转矩 3-8
质量流量 1-7
质量平衡 10-2
秩 20-9
秩和检验法 16-2
智能变送器 12-5
智能传感器 19-1
智能符号 8-1
智能功率模块 4-1
智能化仪器 12-6
智能家居 15-2
智能建筑物管理系统 14-5
智能建筑物系统 14-1
智能卡 14-5
智能控制器 14-5
智能控制系统 12-1
智能楼宇 14-1
智能现场设备 12-5
智能型分散控制器 14-3
智能压力变送器 1-2
智能仪表 11-1
智能仪器 1-1
智能粘贴 17-3
滞后 20-6
滞后环节 20-2
滞后时间常数 18-1
滞后校正 22-4
滞回比较器 22-9
置位 2-2
置信概率 16-2
置信水平 16-2
置信系数 16-2
中点 5-12
中断 2-4
中断标志位 2-5
中断传输 19-3
中断堆栈 6-4
中断返回 2-4
中断方式 2-7
中断服务程序 2-4
中断请求 2-4
中断请求标志 2-5

中断入口地址 2-4
中断设备 12-1
中断条件 2-4
中断响应 2-4
中断溢出 2-5
中断源 2-4
中断允许 2-4
中断组织块 6-4
中规模集成电路 21-4
中继器、网桥、交换机和路由器 9-7
中频 22-4
中线 3-5
中心点 20-10
中心频率 22-2
中性点箝位型逆变电路 4-5
中央处理器 2-1
中央对齐 8-1
中央控制单元 12-1
中央控制室 14-4
中值滤波 12-6
终点 20-4
终端服务 8-1
终端设备 15-3
终端设备绑定 15-4
终值 10-3
终值定理 20-3
终止角度 17-2
终止啮合点 13-6
重力 3-2
重载运行 3-3
周波变换器/交交变频 4-4
周期 10-3
周期采样 12-2
周期性变化的系统误差 16-2
周期性的 21-1
周转轮系 13-6
轴 13-10
轴承 13-9
轴承衬 13-9
轴承承载能力 13-9
轴承径向载荷 13-9
轴承宽度 13-9

轴承内径 13-9
轴承套圈 13-9
轴承外径 13-9
轴承系列(滚动轴承) 13-9
轴承压强 13-9
轴承轴向载荷 13-9
轴端挡圈 13-8
轴肩挡圈 13-8
轴颈 13-9
轴套 13-9
轴瓦 13-9
轴向载荷系数 13-9
逐次逼近 21-10
逐次逼近寄存器 19-2
主磁通 3-5
主从触发器 21-5
主从模式 9-6
主导极点 20-4
主电路 4-1
主调节器 10-4
主动带轮 13-7
主对角线 20-9
主回路 10-4
主机 19-3
主控继电器 6-1
主控制器 19-3
主配线架 14-6
主绕组 3-10
主设备故障时启用的备份服务器 8-1
主要带 20-8
主站接口模块 6-3
住宅产业集成建造系统 14-6
助记符 2-3
铸钢 3-3
铸铁 3-3
专家控制系统 12-1
专家模糊控制 12-9
专用集成电路 21-4
专用接口版 12-4
专用接口芯片 12-4
专用型 22-6
转差功率 18-5
转动副 13-3

索 引

转动惯量 3-2
转换 6-5
转换测量 1-2
转换技术 2-7
转换连线交叉点为连接点 14-6
符号 17-2
转换时间 19-2
转换元件 1-1
转接点 14-6
转矩 3-2
转矩极性鉴别器 18-3
转矩转速曲线 3-8
转速 3-2
转速测量 1-7
转速超调 18-2
转速传感器 1-4
转速调节器 18-2
转速负反馈 18-2
转速计 10-4
转移导纳 5-14
转移函数 5-14
转移特性 22-4
转移阻抗 5-14
转折频率 20-5
转子 3-3
转子导条 3-6
转子堵转 3-6
装配 13-3
装载用户程序 6-4
状态 20-9
状态变量 20-2
状态表 21-6
状态方程 20-9
状态估计 20-9
状态观测器 20-9
状态化简 21-6
状态空间 20-9
状态矢量 20-9
状态图 21-6
状态转换 21-6
状态转移矩阵 20-9
锥角 13-6
锥距 13-6
锥孔轴承 13-9
准互补电路 22-4
准确度 16-1

准时间最优控制 18-2
准稳态 18-2
准谐振 4-7
桌面型会议电视系统
子系统 22-1
紫外线 2-1
字 2-1
字符串 2-3
字符校验 9-4
字节 2-1
自变量 20-10
自补偿 19-5
自持振荡 20-10
自底向上设计 17-2
自顶向下设计 17-2
自动闭锁信号 4-8
自动布线 17-5
自动调节阀/带伺服电机 10-4
自动调心滑动轴承 13-9
自动放置电气连接点 17-2
自动跟踪 12-8
自动化技术 12-5
自动化系统 6-1
自动加料罐 10-4
自动开关 3-8
自动控制 12-7
自动控制牵引设备 4-8
自动控制系统 18-1
自动缩放 17-2
自动校正 12-6
自动整定 12-1
自动执行器 10-2
自动重发控制 12-5
自动状态 10-2
自感 5-10
自感式传感器 1-2
自激振荡 22-7
自检 19-5
自举电路 22-7
自控离合器 13-8
自冷 3-5
自励 3-3

自耦变压器 3-5
自起动 3-10
自然采样法 4-6
自然频率 10-3
自然响应 5-7
自润滑滑动轴承 13-9
自适应控制 12-1
自校准 19-5
自学习 12-9
自由电子 22-3
自由度 13-3
自由度数 13-3
自由分量 5-7
自愈环网 15-2
自诊断 19-5
自主测量仪器 19-1
自组织 15-1
综合布线 14-6
综合布线系统 14-1
综合业务数据网 14-6
综合语音数据终端 14-6
总电流 6-3
总貌 8-1
总线 2-2
总线工业控制机 12-1
总线节点 6-6
总线宽度 17-2
总线连接 6-6
总线请求 12-5
总线拓扑 15-2
总线响应 12-5
总线协议 6-6
总线引入线 17-2
总线仲裁 11-4
总线周期 2-2
阻抗 5-9
阻抗变换器 1-2
阻抗参数 5-14
阻抗参数矩阵 5-14
阻抗的并联 5-9
阻抗的串联 5-9
阻抗基值 3-5
阻抗模 5-9
阻抗三角形 5-9
阻尼 3-10

阻尼比 20-3
阻尼绕组 3-7
阻尼自然频率 20-3
阻容耦合放大电路 22-4
组合 13-3
组合测量 16-5
组合件 13-3
组合逻辑 21-2
组件 22-1
组态 8-1
组态软件 11-1
组态选项 6-2
组态用户系统标识符 8-1
最大超调量 20-3
最大功率传输 5-4
最大过盈 13-10
最大极限尺寸 13-10
最大间隙 13-10
最大实体极限 13-10
最大输出幅值 22-4
最大输出功率 22-8
最大误差法 16-2
最大限度的提高性能 22-1
最大项 21-2
最大转矩 3-7
最低位 21-1
最低位 2-3
最高工作结温 4-1
最高位 2-3
最佳测量方案的确定 16-3
最少拍控制 12-8
最小二乘法 1-1
最小二乘法原理 16-5
最小过盈 13-10
最小极限尺寸 13-10
最小间隙 13-10
最小实体极限 13-10
最小项 21-2
最优控制 12-1
左半平面 20-3
左对齐 8-1
左手定则 3-6
坐标 3-8